Evolution of the

RADIO

Edited by Scott Wood

ISBN# 0-89538-004-8

Copyright 1991

L-W Book Sales
P. O. Box 69
Gas City, IN 46933

Table of Contents

Title Page ....A
Table of Contents ....B
Introduction ....C - D
Display Ads ....E - G
Dealers and Collectors ....H - J
Black & White Catalog Pages and Magazine Ads ....1 - 91
Color Catalog Pages and Magazine Ads.....................92 - 102
Color Photos ....103 - 206
Color Postcards ....207 - 209
Price Guide ....210 - 217

INTRODUCTION
By Bernard Goldman

We think of radio as a forerunner of television by just 25 years or so. However, in reality, Heinrich Hertz, Oliver Lodge, Lee De Forest and Guglielmo Marconi theorized and then proved that sound waves could be transmitted without wires one hundred or so years ago. This indicates that man is or has been able to create anything he has a mind to (and sometimes does).

Radio came into practicality on November 2, 1920. KDKA in Pittsburgh, Pa., by announcing that Warren G. Harding had been elected president. This was the first U.S. licensed station to broadcast (the first is under scrutiny) commercial statements.

Science is great, but collecting scientific accomplishments is even greater! This is the opinion of a prejudiced collector, for no one can resist looking for early crystal sets. These contraptions really were impractical primarily because they required earphones and only one person could listen. It took several years before loud speakers were perfected in order to transmit sound to more than one person at a time. This development then spread commercials, ideas, beliefs, soap operas, mystery shows, quizzes, and a great deal of pleasure to vast audiences.

The impact of early radio was immeasurable for it probably elected a president, Franklin Delano Roosevelt, and defeated his opponent, Herbert Hoover, the incumbent. The distinct voices of Jack Benny, Bing Crosby, Rudy Vallee, Burns and Allen as well as a myriad of other entertainers, were broadcasted to and audience starved for entertainment (particularly during the Depression). It succeeded in making many of them millionaires.

As the 1930s arrived, so did a new era whereby the living room had almost the same impact on a family as the bedroom. Times were tough. Dad was unemployed (most of the time) and going to the movies cost from ten cents to as much as a quarter for just two hours or so of entertainment. With an expenditure of as little as $9.95 for a midget AC-DC table model radio, a family could be supplied with much pleasure for many years at a cost of less than one penny per day.

Now, though classic fancy wood consoles of the Art Deco designs bring mind-boggling prices, it is in the plain wooden table models of the late '20s followed by the 1933 plastic units which created a revolution in the entertainment field, that have lasted almost until this day. They were known and called by most, "midget sets", and have become extremely collectible.

Design-wise, these sets varied from cathedral peaks to rounded peaks, to ridged peaks as well as to square flat tops and to elongated flat tops. Sometimes the wooden table radio appeared with a slant front called "tombstone" that seemed as though its legs had been cut.

Gradually that style disappeared and its fromt stood straight up. Designs remained constant until just about the beginning of World War II, when the late Art Deco era faded into oblivion.

Plastics, such as Bakelite and Catalin, were in common use for many years even though most designers were afraid to compete against wood... the main ingredient in the artistic models of their competitors. It seemed whether the radio was of pressed wood or metal it wasn't plastic, and that is what the manufacturers preferred.

Over the years, who but a collector can remember these *unfamous* manufacturers names: Adler-Royce, Murad (I thought it was a tobacco), Fada, Grebe, Atwater-Kent, Somerset, Sparton (eventually TV), American Bosch, Kadette, Harlson (my family's first), Powertone, Firestone, Belmont, Temple, Majestic and Scott (very classy and expensive).

And then there are names that still remain familiar: Arvin, RCA, Zenith, Motorola, Philco, Silverstone (Sears), General Electric, Crosley (a.k.a. the Model "T" of radio), Emerson (a comparatively inexpensive set that charmed me with its 1933 four-tube Mickey Mouse models), and Westinghouse, etc.

As soon as America recovered from the Great Depression, the average household had more than one radio. Generally speaking, a console was located in the living room, and a table model usually was set up in the kitchen for mon's entertainment. Affluence permitted a many knob and wooden dial set in the master bedroom while a cartoon character model was placed in the kid's room. Perhaps a portable battery operated unit would be kept as a spare of for porch listening.

As it is with most collectibles, condition is paramount in arriving at its value. Since the 1930s, table model radios were mass produced and their initial quality wasn't the greatest. Thus, appearance and performance sets their value. Strangely enough, the "Depression" compact radio of the Roosevelt era seems to most in demand today. In particular, the "cathedral" madels and the 8-tube or more all-wave sets are quite popular also.

Civilian Radio production came almost to a complete halt during World War II due to a scarcity of raw materials scheduled for wartime priorities. The only collectible from that period would be boot-leg or blackmarket sets...rare, but not desirable.

The field of radio collecting is so vast that one needs to specialize in just one aspect, such as table models, to be fulfilled. Also, without the invention of the radio, notwithstanding television's appearance, NBC, CBS and ABC would be just letters in the alphabet.

SO HAPPY HUNTING !!

We would like to thank Bernie for permission to use his introduction in this book. We would also like to thank his wife Gwen who donated many of the catalog pages. They are both fine dealers of Paper Americana and also great people.

Gwen's Antiques
4 Michael Lane
Denver, PA 17517

Format and Pricing Information

In using this book you must remember that information and photos came from many different people from throughout the country. In some cases information may have been a little sketchy and some of the photos low quality. We have tried our very best to include all photos and appreciate all who donated. Most photos are self explanatory so we haven't included information that we felt would be unnecessary. Prices stated in the Price Guide are for radios in fine condition (Not necessarily the radio in the photo). Pricing was based on many factors.... Auction prices, Show prices, Owner's opinions, etc. The prices for the catalog pages and magazine ads are for the radios pictured in them. The prices for the postcard pages are for the postcards. Remember, this is only a guide and should be used as such. L-W Book Sales is not responsible for gains or losses in using this book.

If you would like to have pictures of your radios in the next volume please contact L-W Book Sales for information.

RADIO DEALERS & COLLECTORS

Allen, Jim, 1653 New CAstle Dr., Los Altos, CA 94024

Allen, Sidney, 1908 Rose Crest Dr., Greensboro, NC 27408

Antique Electronic, 6221 So. Maple Ave., Tempe, AZ 85283

Arnold, Gary, 615 Oak St., Marion, NC 28752

Bacon, Fred, 42716 51st St. W., Quartz Hill, CA 93536

Barborak, Gil, 7803 Los Indios Cove, Austin, TX 78729

Bazin, L.J., 27 Shawnee Ln., Vincentown, NJ 08088

Becker, Joan, 10 Mawal Dr., Cedar Grove, NJ 07009

Berg, Jim, 4261 Wilcox Rd., Northport Rd., Northport, WA 99157

Borrelli, Guy, 11441 Francis Green Dr., Gaithersburh, MD 20878

Bray, Charles, 1322 Ivy Rd., Bremerton, WA 98310

Breckenridge, George, Jr., 17790 West Pond Ridge Circle, Gurnee, IL 60031

Breed, Robert, 5110 Los Altos Ct., San Diego, CA 92109

Brown, Dale, 2195 Miner St., Apt. 3, Costa Mesa, CA 92627

Burkhart, Gene, 4528 Stanley Dr., Stephens City, VA 22655

Byrd, Mark, 6710 Sutter Park, Houston, TX 77066

Carrieri, Nicholas, 11 Kunigunda Pl., Islip Terrace, NY, 11752

Carter, Denny, 601 School St., Box 445, Karnak, IL 62956

Chinskey, John, 714 Cagney Ct., Bel Air, MD 21014

Cowlfield, Bill, 17009 Jeanine Pl., Granada Hills, CA 91344

Crismond, Robert, 585 N. Plymouth BL, Los Angeles, CA 90004

Cutler, Dan, P. O. Box 1203, Douglas, WY 82633

Davis, Dick & Nina, 2655 West Park Dr., Baltimore, MD 21207

Daveler, Jay, 5th & Cannon Ave., Lansdale, PA 19446

Doggett, Spencer, 61333 N. Ridge Trail, Romeo, MI 48065

Dressler, Jeff, 161 Allan Dr. #3, Newport News, VA 23602

Duck, Douglas, 3821 Hollow Creek Rd., Ft. Worth, TX 76116

Durand, Michael, 21 Legrande Ave., Tarrytown, NY 10591

Dziedziz, James, 1263 Springer, Westland, MI 48185

Eberley, Hal, 6412 S. 77 Circle, Ralston, NE 68127

Emily, Carol, 1053 W. Outer 21 Rd., Arnold, MO 63010

Evans, Bob, 5722 West North Ave., Milwaukee, WI 53208

Eynon, David, 106 No. Roberts Rd., Bryn Mawr, PA 19010

Feldt, Mike, 12035 Somerset Way East, Carmel, IN 46032

Fender, Gary, Box 639, Escatawpa, MS 39552

Fischer, Richard, Gas City, IN 46933

Fitch, Bill, RD 2, Rt. 23-200, Pottstown, PA 19464

Frantz, F., 100 Osage Ave., Somerdale, NJ 08083

Freeman, George, P. O. Box 369, Carrollton, KY 41008

Fullerton, Kenneth & Sharon, 2341 E. Livingston Ave., Columbus, OH 43209

Fultz, Dick, C/O Richard James Corp., 1209 Donnelly Ave. #203, Burlington, CA 94010

Gerber, W. R., 225 Capitol Rd., Southern Pines, NC 28387

Gilmore, Doug, 1445 Flora, Reedley, CA 93654

Ginocchio, Thomas, 719 Carroll St., Apt 4R, Brooklyn, NY 11215

Goodwin, Nola, Box 768, Springdale, AR 72765

Grabowski, Mike, 3428 So. 88th St., Milwaukee, WI 53227

Graczyk, Ed, 6308 Kiowa N. E., Albuquerque, NM 87110

Grove, Fredi & Len, 4584 W. Ivanhoe ST., Chandler, AZ 85226

Gutzke, Kim, 7134 15th Ave. So., Minneapolis, MN 55423

Hanke, Mike, 1036 S. 15th Ave., Wausau, WI 54401

Harris, 711 E. 44th St., Savanna, GA 31405

Harris, William, 1513 Bellechase Dr., Roanoke, TX 76262

Heimstead, Doug, 1349 Hillcrest Dr., Fridley, MN 55432

Hentges, John, N8012 Washington Dr., Spokane, WA 99208

Hill, Gary, 1507 Ridge Ave., New Castle, PA 16101

Hoffman, Elliot, 3255 Sunset Ave., Eagleville, PA 19403

Hubbard, Johnny, 2999 Osborne Rd. NE, Atlanta, GA 30319

RADIO DEALERS & COLLECTORS

Hyman, Charles, 3804 Wildwood Ct., Monmouth Jct., NJ 08852
Jarrett, John, 201 W. Diamond St., Kendallville, IN 46755
Johnson, J. H., 560 Meadowview Lane, Greenwood, IN 46142
Johnson, Jim, 2635 Ardis, Las Cruces, NM 88001
Johnson, Tom, 215 E. 7th St., Ames, IA 50010
Johnson, Warren/Johnny, 2201 Newton St., Denver, CO 80211
Jonas, Gilbert, 215 E. 80th St., New York, NY 10021
Jones, Keith, 2901 Georgia, Amarillo, TX, 79109
Kay, Julia, 5 Fiske St., Worcester, MA 01602
Keller, Mike & Kathleen,   614-231-3489   need address
Kendall, David, 401 Himes St., Huntington, IN 46750
Kendall, J. E., P. O. Box 436, Fallston, MD 21047
King, Randy, 3017 Orwell, Lincoln, NE 68516
Kinnard, Jay, 623 Amesbury Lane, Austin, TX 78752
Kittleson, Charles, 1944 Oak Knoll Dr., Belmont, CA 94002
Komon, J., Tirschenreuthertr. 4, 8000 Munchen 90, W. Germany
Kroegel, James, 856 Old Farm Rd., Columbus, OH 43213
Krueger, James & Felicia, New Wireless Pioneers, P. O. Box 398, Elma, NY 14059
Lake, Helmut, 295 Hooker Ave., Poughkeepsie, NY 12603
Lambert, Maurice, 1921 Blue Bird Ave., Ft. Worth, TX 76111
Larsen, Gerald, 7841 W. Elmgrove Dr., Elmwood Park, IL 60635
Lasseron, David, 707 Amoroso Pl., Venice, CA 90291
Lauritsen, Robert, 33195 Case St., Lake Elsinore, CA 92330
Leeth, Carol, 801 S. Webster #14, Anaheim, CA 92804
Lewis, Michael, 6070 Lincoln Rd., Oregon, WI, 53575
Lopes, Stanley, 1201 Monument Blvd. #74, Concord, CA 94520
Lupo, Philip, 20-59 47th St., Astoria, NY 11105
Mabbs, Merrill, 709 Pluma Dr., Rapid City, SD 57702
Mac's Old Time Radios, 4335 W. 147th St., Lawndale, CA 90260
Mason, Ross, 715 S. Penn, Mason City, IA 50401
Mathews, Bertha, 2007 Greenland Dr., Murfreesboro, TN 37130
Matthews, Richard, 8212 Timber Trail, Chagrin Falls, OH, 44022
McCoy, Mike & Karen, 8104 NW 114th, Oklahoma City, OK 73162
McElroy, Howard, 2309 Driftwood Apt 1007, Mesquite, TX 75150
McNamara, Terry, 718 1/2 Hammond Ave., Waterloo, IA 50702
Mears, Mary, 4110 S. Troost Pl, Tulsa, OK 74105
Mednick, David, 1450 Palaiside #5H, Firt Lee, NJ 07024
Melvin, Steve, 34 Deep Brook Harbor, Suffield, CT 06078
Michelson, Ralph, 4538 Golfview Dr., Brighton, MI 48116
Mike's Signs, 3981 Nine Mile, Warren, MI 48091
Molettiere, Sylvia, 105 Main St., Souderton, PA 18964
Moore, T. A., 7205 No. Fenwick Ave., Portland, OR 97217
Morey, Mike, 866 So. Bates, Birmingham, MI 48009 313-540-1603
Nordboe, Don, 3220 W. Broadway, Council Bluffs, IA 51501
Nunn, Keith, Rt. 1 Box 136B, Snow Cap, NC 27349
Olawski, Robert, Golden Era Radios, 230 Court Ave., Lyndhurst, NJ 07071
Oppenheim, Peter, 146 E. 49th St., Apt 3A, New Yor, NY 10017
Osborne, Dennis, P. O. Box 5096, Raleigh, NC 27650
Owens, D. K., 478 Sycamore Dr., Circleville, OH 43113
Parks, Anthony, P. O. Box 452, Franklin Grove, IL 61031
Paul, Floyd, 1545 Raymond, Glendale, CA 91201
Pikarz, Bob, 111 E. 29th St., LaGrange Park, IL 60525
Piorek, Alan, 1050 N. Paulina St., Chicago, IL 60622
Pipa, James, 80 Circle Dr. W., Patchogue, NY 11772
Play Things of the Past, 3552 West 105th, Cleveland, OH 44111
Pralskin, Michael, 7235 Hollywood Blvd. #315, Hollywood, CA 90046

Pupo, Gene, 1907 W. Liberty Ave., Spokane, WA 99205
Randy King's Photography by Mary Ann Holland, Lincoln, NE
Rankin, James, 3445 Adaline Dr., Stow, Oh 44224
Reid, George, 1306 Pine Dr., Kalleen, TX 76543
Reynolds, Tom, 5260 McCandlish Rd., Grand Blanc, MI 48439
Richard's Radios of Omaha
Riff, James, 81 N. Ela Rd., Barrington, IL 60010
Rogers, William, 13305 Court Ridge Rd., Midlothian, VA 23112
Round Up Electronics, 2927 NE Riverside Ave., Pendleton, OR 97801
Rouse, Jim, 949 Hibascus Ln., San Jose, CA 95117
Runyon, Bob, 7019 Shawlis Dr., Reynoldsburg, OH 43068
Santoro, Bud, 3715 Bower Rd., Roanoke, VA 24018
Schisler, Vernon, 4144 Wright, Saint Anne, MO 63074
Schneider, Gary, 9511 Sunrise Blvd., #J-23, North Royalton, OH 44133
Schoen, Russell, E. 7340 Nietzke Rd., Clintonville, WI, 54929
Schoenig, Roger, 1946 Washburn, Cincinnati, OH 45223
Speno, Michael, 7-11 E. Genesee St., Auburn, NY 13021
Stoyanoff, Chad, 2940 E. Highland Rd., Highland, MI 48031
Synek, William, 2014 Quail Ridge Dr., Plainsboro, NJ 08536
Thomas, Brad, 3891 Logan Ct., Concord, CA 94519
Turner, J. W. , Rt. 1 Box 58C, Blanch, NC 27212
Vanpragg, Phillip, 0 N 329 Willow Rd., Wheaton, IL 60187
Wagner, ED, 519 E. 1st Ave., Shakopee, MN 55379
Warner, Eugene, 522 Weman, Ridge Crest, CA 93555
Wibbels, Debbie, 5002 Weber Ln., Floyds Knobs, IN 47119
Wiggert, Dave, 1025 Oak Ave. So., B-21, Onalaska, WI 54650
Williams, James, 325 Marlin Rd., White House, TN 37188
Young, James, Rt. 14, Box 593, Florence, AL 35630 205-764-5969

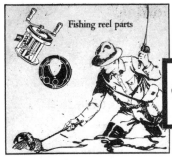
Fishing reel parts

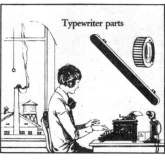

2

RADIO AT ITS *BEST*

Rich Toned   *Single* Tuned   *Self* Contained   Compact

All of these advantages flow from a single source

THE remarkable thing about Pfanstiehl radio is its utter simplicity. From that one merit flow many desirable qualities: beautiful tone, single tuned control, graceful design, dependable operation. It always *works;* is easy to tune, takes up little room.

People no longer care about "stunts" in radio. They want SERVICE. Women especially want uncomplicated tuning and fine musical tone. They want a compact and graceful piece of furniture. They do not want machinery in the home.

In the Pfanstiehl "overtone" receivers there is a close connection between simple tuning and beautiful tone. They have the same scientific source.

The Secret of Radio Tone

The secret of radio tone lies in an uncomplicated circuit, so that the stream of vibrations passes UNHAMPERED through the set, while it is being immensely amplified first in the radio frequency stages and then in the audio frequency stages. This stream of vibrations is ex-

tremely delicate; any distraction or absorption of it at any point distorts the tone. The problem is to keep the stream intact. This Pfanstiehl has accomplished by extremely simple means—a unique achievement in radio design.

He guides and controls the radio stream with very few elements. No extra parts are required to correct amplification errors in the circuit—there are none to correct. There is no distortion of the delicate vibrations which make overtones. Hence the Pfanstiehl is called the "Overtone" receiver.

MODEL 20
"Single-tuned Six" Table Set.
$125
West of the Rockies, $130

1926 Pfanstiehl Catalog Page

MODEL 202

Console Single-tuned Six with inbuilt speaker.
Space for A&B batteries and charger or Socket Power equipment
$210
West of the Rockies, $217.50

1926 Pfanstiehl Catalog Page

4

The Importance of Overtones

The importance of overtones in voice or music needs no stressing to the musical mind. They are the life and soul of beautiful sound. They make what is called tone color and they distinguish between one voice and another, between one musical instrument and another, between one performer and another. They are the medium through which a Paderewski expresses his temperament and genius.

Radio tone is largely a matter of comparison. Many radios are decidedly poor in tone. Others are fairly clear and not displeasing. But compare them with a Pfanstiehl "Overtone" receiver and you will know the difference. Its tone is much richer and purer; has "soul" in it. That is all due to the perfect reproduction of the overtones.

So sensitive and true is Pfanstiehl "Overtone" reception that leading vocal teachers use it to register the breathing impulses of professional singers. Cello and violin teachers use a Pfanstiehl to study the individual tone quality of great performers.

Single Tuned Control

Without Auxiliaries

In the Pfanstiehl "Overtone" six tube receiver there is but one tuning knob to turn. You quickly get whatever station you want, its wave length appearing in the illuminated disc window. As a matter of fact, no knowledge of wave lengths is needed. You can tune by ear, BLINDFOLDED. As you sweep through the scale you catch everything on the air in quick succession.

Consider what a new field of enjoyment is thus available in the home. From the oldest to the youngest, everybody

MODEL 201-A
"Highboy" 18 inches
wide, 74 inches high.
Console model with overtone speaker on top.
Space for A&B batteries and charger or
Socket Power equipment. Six tubes.

$235
West of the Rockies, $247.50

1926 Pfanstiehl Catalog Page

5

can operate the radio and actually get better results than can be had by multiple tuning. The tuning is not only easier and quicker; it is more accurate in the hands of anyone but an expert.

The Pfanstiehl single tuned control was made possible a year ago by the perfect evenness of the radio stream in passing through the various circuits. These are absolutely equal ELECTRICALLY. Hence they can be tuned all at once with a single control, without the aid of any verniers or adjusters to complete the tuning. This year the tuning panel has been still further simplified. It contains just three things: the tuning knob, a combined switch and volume control and the illuminated wave length window. NEVER BEFORE IN THE HISTORY OF RADIO DESIGN HAS SO SIMPLE AND CLEAN-SWEPT A PANEL BEEN OFFERED.

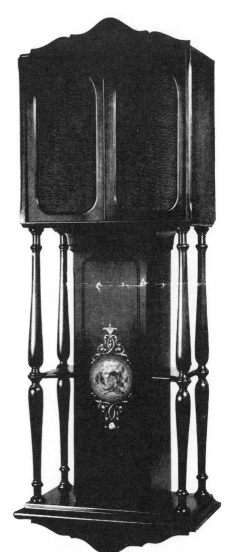

OVERTONE WALL SPEAKER
33 inches high, 12 inches wide.
$65
West of the Rockies, $70

MODEL 201
"Single-tuned Six" Console without speaker, but space for A&B batteries and charger or Socket Power equipment. Six tubes.
$170
West of the Rockies, $177.50

There are three types of radio sets on the market, each suitable for its own purpose. There is the ordinary set giving ordinary service. It is not a fine musical instrument, you use it like a telephone to get information over the air. There is the highly complicated set which does "stunts" in the hands of an expert and under favorable atmospheric conditions. And there is the type of set of which Pfanstiehl is an outstanding example, rich toned, simply operated, always the same; also gracefully designed and beautifully made. The Pfanstiehl appeals to refined home folks who love beauty, whether in tone, line or color. At the same time it has all the selectivity, volume, and reach desirable, without sacrifice of important qualities, such as tone, dependability, and simple operation.

1926 Pfanstiehl Catalog Page

SOMETHING REALLY *NEW* IN CABINET DESIGN

Self-Contained   Unique   Compact

There is so little machinery in a Pfanstiehl, inside and out, that an extremely compact radio cabinet is possible. It is only twenty inches wide. The console model is even narrower and only forty-one inches high. And still it has room for every accessory needed to operate a radio: inbuilt speaker, A and B batteries and charger, or Socket Power equipment. This makes a "petite" piece of furniture for the home; and it makes a nicely balanced and perfectly adjusted radio unit. The console is a space saver. It fits anywhere—between windows, in a corner, between doors. It can be conveniently placed in living room, in bed room, in den or sun parlor, HOWEVER CONTRACTED THE SPACE. It may be had with or without inbuilt speaker.

An attractive feature of the console is its metal grill opening for the speaker—of ornamental design and antique gold finish.

The "Highboy"

This is a more elaborate console model eighteen inches wide, with a tall, graceful superstructure for the speaker, extending six feet from the floor. There is nothing like it in radio design. It resembles a grandfather's clock, or perhaps more nearly the tall, slender china cupboard of old English design. It comes in hand-rubbed walnut finish or in Chinese lacquer red, blue or yellow.

The Wall Speaker

An unique speaker has been designed for use on top of the console or to be hung on the wall if you prefer. It very much resembles an old fashioned wall clock, flanked by four graceful spindles and supported by a narrow shelf at the bottom. From the standpoint of acoustics there is a real advantage in having the speaker mouth six feet or more from the floor. The tone vibrations are better distributed. The Pfanstiehl speaker is an overtone instrument sympathetically adjusted to the Pfanstiehl overtone re-

MODEL 18
Single Dial Dual Control Table Set. Five tubes.
$95
West of the Rockies. $100

1926 Pfanstiehl Catalog Page

7

MODEL 181
*Single Dial Dual Control Console
without speaker, but place for A&B
batteries and charger or Socket Power equipment*
$135
West of the Rockies, $142.50

A Special Single Dial, Dual Control, Model

For people who have less money to invest than the single tuned model costs, Pfanstiehl has developed a new dual control receiver which is very unique. One tuning knob controls a master pointer indicating wave lengths on the station chart. The other controls a smaller pointer operating within the master pointer. This is in no sense a two-dial model. It has but one dial and two concentric pointers. The eye has but one place to watch in scanning wave lengths.

This model has five tubes; is well made and handsomely finished.

ceiver. The result is a beauty of tone which is extraordinary.

A Fine Piece of Furniture — Not a Machine

In designing these new cabinets it has been the Pfanstiehl aim not only to make them distinctive, but also as simple, as compact and as graceful as possible and thus fit radio into the atmosphere of a fine home. The impression is not so much that of a machine as of furniture. The panel is of hand-rubbed walnut, with wooden knobs—a handsome combination of rich brown and old gold.

MODEL 182
*Single Dial Dual Control Console
with inbuilt speaker. Space for A&B batteries
and charger or Socket Power equipment. Five tubes.*
$170
West of the Rockies. $177.50

1926 Pfanstiehl Catalog Page

1925 Magazine Ad

1925 Magazine Ad

1925 Magazine Ad

1925 Magazine Ad

Six Radio Sets
for the man who believes his own ears!

Merry Christmas

ALL the resources, effort and experience that go to develop a radio set can be verified and measured with one question—"How do your own ears like it?"

A-C DAYTON offers many refinements—it offers Second Stage Tuning, Radio's greatest refinement—but these are important only because they will *please* your ears, because they will make you *like* A-C DAYTON reception better than any radio you have ever heard.

You can make your choice from six models and six prices. It will always be an A-C DAYTON.

Trust your own ability to judge good radio. When you have heard other fine receivers go to the store that sells A-C DAYTON and let each set speak for itself. Notice how you can use special tuning controls to bring you nearer to your favorite stations and to overcome those local conditions which will interfere with ordinary sets.

Important A-C DAYTON Refinements

Double Vernier Dial Control
Air-Spaced Coils
Double Reading Voltmeter
Phosphoric Reproduction
Completely Shielded Coils
Cushioned Tube Sockets
Sensitivity Control Compensator
Fully Graduated Volume

THE A-C ELECTRICAL MFG. COMPANY, DAYTON, OHIO
Makers of Electrical Devices for More Than Twenty Years

The True Proof of Radio
For six years A-C DAYTON has been the radio for the man who believes his own ears. This advertisement can tell you a little; your own ears can tell you everything. Make that your final test. If you do not know the A-C DAYTON dealer near you send us the coupon today.

THE A-C ELECTRICAL MFG. CO.
DAYTON, OHIO
Gentlemen:
Without obligation send me full information on A-C DAYTON Receivers and name of nearest store which makes demonstration tests.
Name
Address

For the man who believes his own ears

A-C DAYTON
RADIO

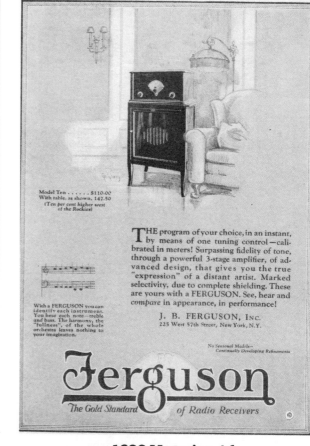

Model Ten $110.00
With table, as shown, 147.50
(Ten per cent higher west of the Rockies)

THE program of your choice, in an instant, by means of one tuning control—calibrated in meters! Surpassing fidelity of tone, through a powerful 3-stage amplifier, of advanced design, that gives you the true "expression" of a distant artist. Marked selectivity, due to complete shielding. These are yours with a FERGUSON. See, hear and *compare* in appearance, in performance!

With a FERGUSON you can identify each instrument. You hear each note—treble and bass. The harmony—the "fullness", of the whole orchestra leaves nothing to your imagination.

J. B. FERGUSON, INC.
225 West 57th Street, New York, N.Y.

No Seasonal Models— Continually Developing Refinements

Ferguson
The Gold Standard of Radio Receivers

You hear *all* the tones
with an

ALL-AMERICAN
Reproducer

An All-American Quality Product

A *good* speaker is the only kind worth having. A poor one will ruin otherwise good reception.

We're making a good one for you—the *Lorel* Reproducer; a cone type correctly balanced with sounding-board and sounding-chamber, to give you that purity of *all* tones, which you desire.

This remarkable unit combines the good features of both cone and sounding-chamber types of speaker; and eliminates their inherent weaknesses. You can hear *all* the high and low tones with the *Lorel*; clear and full.

Ask your dealer for a demonstration of the Lorel. You'll find it a real improvement in radio reception.

Price $25 *Slightly higher west of the Rockies*

ALL-AMERICAN RADIO CORPORATION
4205 Belmont Avenue · Chicago

A Remarkable Improvement in Audio Amplification

A development by All-American laboratories—the Rauland-Lyric-Trio. You know the Rauland Lyric Transformer, famous among music critics for its exceptional tone perfection. It is now combined with two Rauland Trio impedance units; retaining the advantages and eliminating the weaknesses of the two leading systems of audio amplification. The result is the last word in audio amplification. *Free book, "Modern Audio Amplification," tells more about this interesting development. Write for handbook "B-90."*

Constant-B
ALL-AMERICAN
PERMANENT PLATE POWER

Pure full tone is possible only with unvarying "B" power. With All-American "Constant B" you get a permanent, constant plate power. There's nothing to take care of; no annoying hum, and no acid. Permanently sealed; "Constant B" has a 20 to 60 volt tap, varied in output by a "detector" control; a 67½ volt and a 90 volt tap; a variable voltage "power-tube" tap uniformly controlled by a "High-Low" switch.

Price $37.50 *Complete with Raytheon tube Slightly higher west of the Rockies*

Rauland=Lyric=Trio

The Amplion Patrician encloses a remarkable 48″ air column, in a graceful, richly carved mahogany cabinet, 18″ x 12″ x 9″. Acoustically it is non-directional, with a new, softly diffused mellowness of tone that makes this instrument the choice of the connoisseur, wherever heard.
AA 18 . . . . $45.00

The new Amplion Patrician reproduces
the very soul of music

—exceptionally rich in those delicate overtones that give to music its temperament, its true character, its tonal color, its sensitive appeal to the spirit.

YOU may own the most expensive radio receiving set. You may tune in on the best radio concerts. Yet, if your reproducer is not delicately and accurately constructed, you will lose most of the *fine* overtones that create the true beauty—the very soul—of music.

Since 1887, engineering experts of "The House of Graham"—the creators of Amplions—have been achieving constant improvement in sound-reproducing devices. As the result of this long experience, it is not extraordinary that the Amplion instruments will reproduce more of music's *fine* overtones, and a wider musical range, than other reproducers are able to do.

AMPLION CONE
Artistically, this new Amplion Cabinet Cone graces the most exquisitely appointed room; of two-tone mahogany, 14″x14″x9″. Acoustically, it is a time perfected Amplion development.
AC12 . . . . $30

AMPLION DRAGON
This model is the best known of all the famous "Dragon" type of Amplions; adopted as standard by leading radio engineers wherever broadcasting exists. Notable for acute sensitivity and amazing volume.
AR19 . . . . $42.50
Other Dragon models from $12 up

Write for the interesting "1927-Amplion" Booklet

THE AMPLION CORPORATION OF AMERICA
Suite AA-1, 280 Madison Avenue, New York City
THE AMPLION CORPORATION OF CANADA Ltd., Toronto

"The House of Graham"—Alfred Graham & Co., of London, England —is known throughout the world through its associated companies.

AMPLION

ALL you want in a radio plus one thing more!

Name what you will that is newest and best in radio. Kolster has it! Remote control—which enables you to operate your Kolster from any remote corner in the home. Electrical tuning—merely pressing a button selects your favorite station electrically. Selector tuner—the new, convenient Kolster method of tuning. Screen grid tubes—for greater distance and quieter reception. Four tuned circuits to choose unfailingly the one station you wish to hear, excluding all else. New dynamic reproducers affording a new realism of tone. Cabinets of extraordinary appeal. And in addition to all these advantages there is the background of fine quality and lasting satisfaction inherent in every Kolster Radio.

Enjoy the Kolster Program every Wednesday Evening at 10 P. M., Eastern Daylight Saving Time, over the nation-wide Columbia Chain.

K-43—Handsome cabinet with doors front panels. Seven tubes and rectifier. Selector tuner. Dynamic reproducer. Screen grid tubes. Push-pull amplification with 2 type 345 tubes. *Price, less tubes* **$235.00**

Model K-45 uses remote control, a Kolster development. With this device, you can steer the radio—take your choice of 8 different stations—and make the volume louder or softer as you will from a remote point in your home!

K-44—Console receiver using 7 tubes and 2 rectifier tubes. Walnut cabinet of tasteful design. Selector tuner of embossed bronze. Screen grid tubes. Dynamic reproducer. Push-pull amplification, 2 type 345 tubes. *Price, less tubes.* **$325.00**

K-45—Richly grained walnut cabinet appearance. Remote control. Electrical tuning. Nine tubes and two rectifiers. Screen grid R.F. tubes. Extra large dynamic reproducer. Three stages of audio . . . second and third stages push-pull, using type 327 tubes and type 350. *Price, less tubes.* **$500.00**

Prices slightly higher west of the Rockies

KOLSTER RADIO

Copyright by Kolster Radio Corporation, 1929

To lend added distinction to the most distinctively furnished room . . . . to thrill the listener with its unsurpassed and unforgettable cathedral tone; that is the unique achievement of Kellogg Radio. And that is why it is found in so many homes of refinement. Kellogg Radio has 3 models of appealing design, including a combination radio and phonograph. All models have automatic volume control, screen-grid tubes, super power tubes and the Kellogg tone-balanced dynamic speaker. Prices range from $250 to $395, not including tubes. Slightly higher on the Pacific Coast.

The RADIO with the CATHEDRAL tone

KELLOGG
SCREEN-GRID RADIO

CORRECTLY ENGINEERED TO THE NEW SCREEN-GRID TUBES

BOSCH • RADIO

In a period when technicalities are being stressed, Bosch Radio again stands foremost in construction details which assure an unrivaled quality of reception. The new Bosch Radio Model 48 is correctly engineered to the new four-element, Screen-Grid tubes. Three are used. There are two type 245 amplifiers arranged in push-pull, powered detector and a type 280 full wave rectifier—seven tubes in all. Only by hearing and operating the new Bosch Radio can you appreciate its greater sensitivity, razor-edge selectivity and tonal accuracy. New artistry in cabinet designing combines beauty with enjoyment. The Bosch Dealer near you will gladly explain the engineering superiority of Bosch Radio. Library model illustrated with sliding doors, less tubes $119.50.

Prices slightly higher west of Rockies and in Canada—Bosch Radio is licensed under patents and applications of R. C. A. R. F. L. and Latelephone.

Bosch Radio Combination Receiver and Speaker Console. Screen-Grid quality is an inexpensive combination of charming individuality. Chosen woods and veneers with rich carving. It has electro dynamic type speaker. Less tubes, $168.50.

New Bosch Radio De Luxe Console the last word in radio—uses seven tubes, three are Screen-Grid. The console has sliding doors concealing dial and electro dynamic type speaker. Antique finish in Old English line. Less tubes, $249.00.

AMERICAN BOSCH MAGNETO CORPORATION—SPRINGFIELD, MASS.

Branches: New York   Chicago   Detroit   San Francisco   You'll Have Better Results With Bosch Radio Tubes

---

Now come living personalities
OVER THE AIR! ▾ ▾ ▾
Through delicate shades of musical feeling never before reproduced in radio

An Amazing Improvement that increases your enjoyment of every type of music

Unbelievable at first—just as Radio was—there comes today a new way of reproducing music. Through a new Stewart-Warner discovery we feel now as never before in radio the influence of the artist's personality.

We feel the cultural effect of music that authorities agree is the greatest known force in our emotional lives.

Yet only when music is absolutely true and faithful to the original can it give these effects. Old methods of radio reproduction could never do it.

So Stewart-Warner set out to build a radio on utterly new principles. We spent 5 years in research. We made and tested 141 designs.

Old ways of controlling outside noise are replaced by a discovery that even eliminates sounds more delicate than the human ear can distinguish. A far more sensitive mechanism is the result. And entirely free from outside noise.

Thus every station you tune to comes in under perfect conditions. You hear music many times more refined than of old. You hear more than sound. You hear the delicate overtones. You hear the soul of the music. You actually live the emotional experience that the artist is translating to you in music.

For a thrill unlike any you have known—hear this new and improved Stewart-Warner Radio—The Set with the Punch. Hear the Punch that brings distant stations booming in with all the strength of a local. Hear the Punch that gives volume, selectivity and realism of tone as never before. See the nearest Stewart-Warner dealer for this new delight.

STEWART-WARNER CORPORATION
Chicago, Illinois, U.S.A.

"THE SET WITH THE PUNCH"

17th Century English Period Console

Approved cabinet No. 47. Of antique finished American walnut, with carved ornamentation. Sliding doors conceal back. With the new Stewart-Warner A.C. Screen-Grid Receiver and built-in Stewart-Warner Electro-Dynamic Reproducer. Built-in aerial—plug-in for television—plug-in for phonograph. All radio and 25 cycle A.C. and D.C. models.

$154.50
less tubes

Complete line of consoles, table models and combinations. $89.50 to $365.50 (less tubes). $40 extra slightly higher west of Rockies.

Stewart-Warner Approved Console Cabinets supplied by Louis Hanson Company, Chicago; Herron Phonograph Corp., Los Angeles; and St. Louis Furniture Co., Stratford, Canada.

Front view sliding doors open and closed

STEWART-WARNER
RADIO
Screen Grid Circuit or Balanced Bridge Circuit

12

BARGAINS IN CONSOLES

Every one of these fine consoles are made by the foremost furniture houses in America. **You cannot buy better** erchandise at any price.

All consoles shown on this page are brand new in original cases and sold at prices never attempted heretofore. asmuch as the supply is limited, we reserve the right to return deposits if items have been sold out. Every one is lly guaranteed, although all consoles shown here are sold far below actual manufacturing cost.

Atwater Kent Approved Dover Model 133 Console

1931 Catalog Page

This console must be seen to be appreciated. The wood alone costs more than we ask for the finished article.

Made from selected Veneers of Walnut with a two-tone routing and other attractive decorations on the front with a heavy full-bodied, hand-rubbed and polished varnish finish. Size: top 16x25½ in., height 38 in., width 24¾ in.

Shipping weight 70 lbs. Packed in original crate.

List Price, $65.00.
No. 133 A.K. Dover Console. **$9.95**
YOUR SPECIAL PRICE

SPECIAL!
With Peerless Type Speaker
Cavalier Model 159 Console

This beautiful and distinctive cabinet is sold with the built-in Peerless Type Speaker. It is designed to provide a maximum cabinet at a minimum price. Veneers of Figured Walnut with attractive carvings and unusual turnings. Bureau door closes and conceals radio panel. Set Compartment, 21¾ in. wide x 10½ in. deep x 8 in. high. Height 38 in.

Shipping weight 65 lbs. Packed in original crate. Complete with speaker.

List Price, $62.50.
No. 159 Cavalier Console. **$9.95**
YOUR SPECIAL PRICE .

Kreeg-Mellen, Model 22 Console

One of our finest consoles.

Walnut finish, Maple overlay. Front door drops. Solid panel for receiver. 21½ in. wide; 38 in. high; 15 in. deep. Finely-hinged bureau top panel drops down when you wish to turn on the radio. Grille below conceals your loud-speaker.

Shipping weight 70 lbs.

List Price, $59.00.
No. 22 Kreeg-Mellen Console **$9.75**
YOUR SPECIAL PRICE

SEE PAGE 2 FOR TERMS

Berkey & Gay, Model 106 Console

A masterpiece by the famous master craftsmen. Oval shown conceals your loudspeaker.

Console opens by dropping well balanced top bureau panel supported by strong hinges.

Walnut, Oak, Bird's-eye Maple and Gumwood combination. 26 in. wide; 38½ in. high; 16½ in. deep. Finest hand-rubbed finish.

Shipping weight 70 lbs.

List Price, $55.00.
No. 106 B & G Console. **$10.90**
YOUR SPECIAL PRICE.

Kreeg-Mellen, Model 1 Console

A REAL BEAUTY!
A CONSOLE THE FINEST HOME WILL BE PROUD OF.

Finely hinged top bureau panel drops down when you wish to turn on the radio. The grille work in lower section conceals your loudspeaker which you place behind the covered grille.

Walnut, Oak, Bird's-eye Maple and Gumwood combination, 21½ in. wide; 38 in. high; 16½ in. deep. Finest hand-rubbed finish. Shipping weight 70 lbs.

List Price $59.00.
No. 111 Kreeg-Mellen Console. **$9.75**
YOUR SPECIAL PRICE

Freed-Eisemann NR-5
(Neutrodyne)

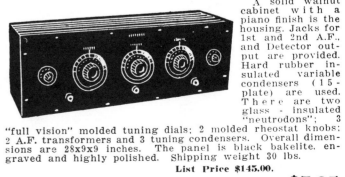

A solid walnut cabinet with a piano finish is the housing. Jacks for 1st and 2nd A.F., and Detector output are provided. Hard rubber insulated variable condensers (15-plate) are used. There are two glass - insulated "neutrodons"; 3 "full vision" molded tuning dials; 2 molded rheostat knobs; 2 A.F. transformers and 3 tuning condensers. Overall dimensions are 28x9x9 inches. The panel is black bakelite, engraved and highly polished. Shipping weight 30 lbs.

List Price $145.00.
No. 2111 Freed-Eisemann NR-5. **$7.95**
YOUR SPECIAL PRICE

$3.00 for the best suggestion. HAVE YOU SEEN IT? (See Page 2).

---

24   **RADIO TRADING COMPANY, 25 WEST BROADWAY, NEW YORK, N. Y.**

Atwater Kent Model 35

One of the most compact receivers ever offered to the public. 3 stages R.F. 3 variable condensers are used. Overall dimensions are: 17½ x 8 x 5½ inches. The chassis is housed in a brown crackle-finish pressed metal cabinet. This is a "one - dial control" receiver. Incorporated in this set is a 6-wire cable, each wire of which is rubber insulated and "color coded." This shielded receiver has very high "gain" and may be used with antennas of any length, without in the least affecting the tuning. The variable condensers are of the "single bearing rotor" type. This set takes the following tubes: 5 type-201A and one type-112A or 171A tubes. Shipping weight 16 lbs. **List Price $65.00.**
No. 2104 A.K. Model 35. YOUR SPECIAL PRICE.. $14.95

Freshman "Masterpiece" A

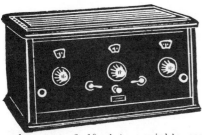

It is of the tuned Radio Frequency type. Requires 4 201A, 1 171A tubes. 2 A.F. transformers, and 3 variable condensers. Overall dimensions are: 20½ x 12 x 9¼ inches; mahogany bakelite panel. The cabinet is finished in mahogany. 3 19-plate variable condensers used. The dial settings are read through recessed windows. 2 jacks mounted on panel. Shipping weight 25 lbs. **List Price, $80.00.**

No. 2101 Freshman Masterpiece. YOUR PRICE $7.00

Freshman "Masterpiece" B

It is finished in grained walnut to match the veneer of the cabinet. Tapered pressed-metal knobs through a 9-to-1 ratio gear control the tuning condensers, the dial setting being observed through windows. Volume, as well as regeneration and oscillation, is adjusted by the two "levers," which are rheostat - arm control. An off-on switch completes the panel layout. Set uses 5 201-A tubes. Two type "30" A.F. transformers are used, ratio of about 5 to 1. Shipping weight 30 lbs. **List Price $75.00.**

No. 2110 Stromberg-Carlson. YOUR PRICE $6.95

Stromberg-Carlson 523

This fine set uses 4 201A and 1 200A tubes. The cabinet is one of the finest ever made for radio sets. A slanting, beautifully grained wooden panel carries the tuning escutcheons. The panel controls include a "Long-Short Antenna" switch; 3-ohm and 20-ohm rheostats; "On-Off" snap switch; audio output jack; and a Weston 0-7 volt-meter. The jack on the panel is for phonograph pickup. A neutrodyne circuit is used. 26 long x 14 deep x 13 inches high. Shipping weight 75 lbs. **List Price $160.00.**
No. 2110 Stromberg-Carlson YOUR PRICE $24.95

Radiola 25 Superheterodyne

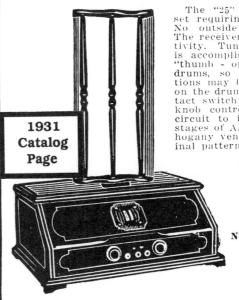

The "25" is a loop-operated set requiring 6 "X-199" tubes. No outside aerial is needed. The receiver has "10-kc" selectivity. Tuning of this receiver is accomplished through large "thumb - operated" tuning drums, so designed that stations may be "logged" directly on the drums. The small center tact switch which changes the knob controls a multiple-concircuit to include one or two stages of A.F. A two-tone mahogany veneer cabinet of original pattern houses the chassis and batteries. Comes complete with loop aerial. Its overall dimensions are 28x19x12 inches high. Shipping weight, 55 lbs. **List Price, $265.00.**

No. 2109 Radiola 25.

YOUR SPECIAL PRICE

$10.95

Ware Type T Neutrodyne

This is the most economical in operation of all radio sets. The circuit is that of a REFLEXED NEUTRODYNE incorporating 3 UV-199 tubes. The mahogany cabinet is 14 in. long and 13 in. deep. This design provides room for the "A" supply of 3 dry cells, 2 "B" and 1 "C" battery. There are 2 15-plate variable condensers, 2 neutrodyne-type R.F. transformers, 2 A.F. transformers, rheostat, 2 jacks, R.F. choke, 2 tuning dials. shock-absorbing mounting for the 3 tubes. Shipping weight 16 lbs. **List Price $65.00.**

No. 2108 Ware Type T. YOUR SPECIAL PRICE $5.95

The Radiola 20

Two stages of tuned frequency amplification, a regenerative detector, and two stages of A.F. amplification, using 4 type X-199 tubes and a X-120 for the last audio stage, is the arrangement of this receiver. The A.F. transformers used in this set are perfectly designed for the required performance Heavy, soft iron encases the windings, and the frequency characteristic is exceptionally good. Cabinet is mahogany. Overall dimensions are: 19x16x11 inches high. Shipping weight 35 lbs.

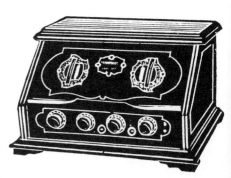

List Price, $102.50.

No. 2102 Radiola 20. YOUR SPECIAL PRICE.. $12.50

RADIO TRADING COMPANY, 25 WEST BROADWAY, NEW YORK, N. Y. 23

Sentinel RADIO

ATTRACTIVE LOW-BOY
7 TUBES—VARIABLE-MU

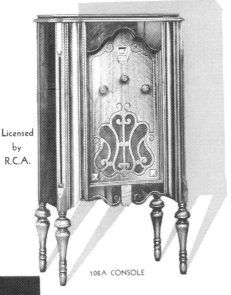

Licensed
by
R.C.A.

106A CONSOLE

RETAIL PRICE 69.50
Complete with Tubes

Dealer's Discount 40%

Dealer's Price
41.70 Less **40.87**
2% Net . . .

Complete with Tubes

This set includes seven matched R. C. A. LICENSED TUBES. There are two 224 screen grid, one 227 oscillator, two 235 or 551 variable-mu, one 280 full wave rectifier and one 247 power pentode tube.

Operates only on 110 volt, 60 cycle alternating current.

TONE QUALITY · SELECTIVITY

1931 Catalog Page

Sentinel RADIO

4 TUBE MIDGET
PENTODE—VARIABLE-MU

The Radio Value of the Age

The Sentinel No. 111 Midget is really a Midget in the true sense of the word. Small enough to fit in most anywhere—only 14¾" high, 10¾" wide and 8½" deep. But what a gem for performance! Sharp and sensitive, it is capable of pulling in station after station with amazing fidelity and volume. And yet, look at those prices! Just think of what sales possibilities are presented here. A four tube set that can be taken from place to place like an ordinary phonograph—yet having such features in design as screen grid, pentode and variable-mu. Make it a point to have one of these tiny marvels to demonstrate and see what wonderful results it will produce in terms of sales and profit.

Engineered for Power

Every new development in radio engineering has been utilized in the design of this little fellow in order that its small size would not stand in efficiency in performance. Good selectivity is obtained through the use of three tuned circuits consisting of a tuned antenna stage, a tuned coupled type pre-selector, tuned detector coil, and the tuned 235 tube plate circuit. A special adaptation of the screen grid tube as a detector, plus the units coupled high uniform gain radio frequency amplification stage and the proper utilization of the pentode tube, is responsible for exceptional sensitivity. Cross talk modulation distortion and power line alternating current modulation, are permanently eliminated by using the variable-mu tube in conjunction with the tuned antenna and coupled type pre-selector. Fully licensed by R.C.A.

Fine Construction

The materials that have gone into the construction of the No. 111 Chassis, have been chosen with a view towards durability. The chassis is made of heavy gauge steel, and cadmium plated. All parts are properly shielded and assembled with utmost precision. Full Vision dial is calibrated in kilocycles.

Sparklingly Beautiful Cabinet

The cabinet although small in nature, is charmingly attractive in its conservative Gothic design. Not too "gingerbready" but with just enough ornamentation to give it a warm appeal. It is made of selected walnut with maple overlays at the sides for high lighting. The speaker grille design is in keeping with the general mold and is beautifully set off by a gold grille cloth.

Flawless Speaker Construction

Every refinement in speaker design found in its big brothers is also incorporated in the speaker of this midget. Sweet, natural, and clear, all sound is reproduced with volume to spare.

Tubes

This set comes complete with four matched R.C.A. LICENSED TUBES. These are one 224 screen grid, one 235 variable-mu, one 247 pentode and one 280 full wave rectifier.
Operates on 110 volt, 60 cycle alternating current only.

RETAIL PRICE 34.50
Complete with Tubes

Dealer's Price 19.75
Less 2%, Net 19.35 ea.

Lots of 3
18.75 Each **18.37**
Less 2% . . . Net

Complete with Tubes

TONE QUALITY · SELECTIVITY

1931 Catalog Page

Sentinel RADIO

SUPERBLY DESIGNED SUP
9 TUBES—VARIABLE-MU

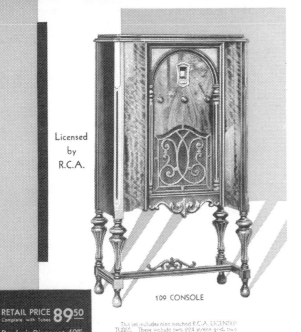

Licensed
by
R.C.A.

109 CONSOLE

RETAIL PRICE 89.50
Complete with Tubes

Dealer's Discount 40%

Dealer's Price
53.70 Less **52.63**
2% Net . . .

Complete with Tubes

This set includes nine matched R.C.A. LICENSED TUBES. These include two 224 screen grid, two 235 variable-mu, two 247 power tubes, one 227 oscillator, one 280 full wave rectifier and one 1110 voltage regulator tube. Operates on 110 volt, 60 cycle alternating current only.

TONE QUALITY · SELECTIVITY

1931 Catalog Page

Sentinel RADIO

An Amazing VALUE—Tabl
7 TUBES—VARIABLE-MU

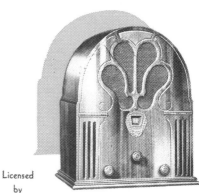

Licensed
by
R.C.A.

108A TABLE MODEL

RETAIL PRICE 59.50
Complete with Tubes

Dealer's Discount 40%

Dealer's Price
35.70 Less **34.99**
2% Net . . .

Complete with Tubes

This set includes seven matched R.C.A. LICENSED TUBES. These are two 224 screen grid, one 227 oscillator, two 235 variable-mu, one 280 full wave rectifier and one 247 power pentode tube.

Operates on 110 volt, 60 cycle alternating current only.

TONE QUALITY · SELECTIVITY

1931 Catalog Page

Clearance Prices on Remaining Stock of Genuine KOLSTER Products

Every item is brand new and in original packing. We honestly believe that these items are the best money makers for wide-awake dealers and represent the finest bargains obtainable. Licensed by R. C. A and Lektophone Patents.

KOLSTER

A. C. DYNAMIC SPEAKER AND POWER AMPLIFIER

Model K5
11½ x 11½ x 19″
Shipping Wght. 75 lbs.
LIST PRICE
$175.00
(without tubes)

This finely matched, rugged unit comprises a complete heavy-duty Electro-Dynamic Speaker, including a 210 or 250 Power Amplifier with "B" supply unit, all self-contained on a steel frame. If desired, the 210 or 250 Power Amplifier will also supply 22, 67 and 90 volts "B" current, sufficient for any set using up to eight tubes. These voltages are brought out to binding posts, conveniently.

Following tubes are required for its operation: 2 UX281 for full wave rectification, 1 UX210 (or UX250) as a Power Amplifier. A 20 foot cable is included with each instrument. The Kolster 10″ Dynamic Speaker has a remarkable tone and does not rattle under power. It is built in a heavy cast-iron case and can stand abuse.

Operates on 110 volts, 60 cycle A. C.

The parts in this unit are worth many times the price you pay and can be used in high power work in constructing an amateur transmitter or amplifier.

$9.50
LESS TUBES
Cat. No. S-3945

1932 Catalog Page

Lowboy Cabinet for Speaker
Shipping Weight 101 lbs.
$2.75
Cat. No. S-3947

KOLSTER

9 TUBE CONSOLE RECEIVER WITH KOLSTER K-5 A. C. DYNAMIC SPEAKER

Model 6H
53 x 27 x 18½″
Shipping Wght. 245 lbs.
LIST PRICE
$295.00
(without tubes)

This receiver, licensed by R. C. A., employs the famous Kolster 6 T. R. F. circuit and operates on either an indoor or outdoor antenna. It uses three stages of R. F., Detector (4 gang condensers) and two stages of A. F. in addition to the 210 or 250 Power Stage in the K-5 unit. A three point tap switch aerial adjuster operated from the panel gives hairline selectivity. By simply adding an "A" eliminator and a small 4½ volt "C" battery, this set can be operated from 110 volts, 60 cycle A. C.

Single dial control, simple to operate.

This receiver embodies everything looked for in a modern set. The Kolster 10″ cone Dynamic Speaker has a remarkable tone and is one of the few which uses genuine leather in its construction. The complete receiver uses 6 UX201A tubes, 2 UX-281 tubes, 1 UX210 (or UX250) tube.

The cabinet is a beautiful highboy console of burled walnut with maple overlay and has full swinging doors.

$18.50
LESS TUBES
Cat. No. S-3946

Cabinet only
Shipping Weight 145 lbs.
$5.90
Cat. No. S-3948

KOLSTER 6-D
SIX-TUBE BATTERY OPERATED RECEIVER

This famous 6-tube battery receiver is the **Kolster 6-D.** For the person who wants a battery-operated receiver, this Genuine Kolster set represents the acme of design and construction. The four-gang condenser, resistors, coils and audio transformers are all of high grade. For selectivity and distance-getting qualities, it is unsurpassed. Using five 201-A tubes in the three R. F. stages, detector and first audio (a UX200-A can be used in the detector to good advantage) with a UX112-A or 171-A in the second audio, this receiver supplies sufficient volume.

Panel: 9 x 15 x 10″
Net Weight: 16 lbs.

Cat. No. S-3949 Table Model **$11.50** (less tubes)   Cat. No. S-3950 Chassis only **$9.50** (less tubes)

Table of Contents on Page "A" of Green Section—Corrections and Additions on Page "D" of Green Section

American
5 Tube Midget Receiver
Variable-Mu Pentode

Uses two 235, one 224, one 247, one 280 tubes.

3 gang condenser (3 tuned circuits)

Rola 6" Dynamic Speaker. Beautiful Walnut Cabinet. Size: 12½" x 15½" x 10". Full Vision Dial. Phono attachment. Volume control. Selective, Sensitive, fine quality with Quiet operation.

For 110 volt, 60 cycle A. C. **$15.75**
Cat. No. 1394

Above chassis and speaker only **$14.50**
Cat. No. 1395

For 110 volt, 25 cycle A. C., with **$18.75** cabinet
Cat. No. 1396

110 volt, 25 cycle chassis and **$17.50** speaker only
Cat. No. 1397

For 220 volt, 60 cycle A. C., with **$17.75** cabinet
Cat. No. 1398

220 volt, 60 cycle A. C. chassis **$16.50** and speaker only
Cat. No. 1399

American
6 Tube Midget Receiver
Variable Mu, Push-Pull Pentode

Uses two 235, one 224, two 247, one 280 tubes.

3 gang Condenser (3 tuned circuits): Tone Control; Volume Control, Phono attachment with "Phono-Radio" switch. Full vision Dial with traveling Pilot Light and indicator. Beautiful Walnut Cabinet. Size: 15 x 19 x 10½". Rola 8" Dynamic Speaker. Quiet operation, selective, sensitive, with ample volume.

For 110 volt, 60 cycle A. C. operation **$21.50**
Cat. No. 1386

Above chassis and speaker only **$19.00**
Cat. No. 1387

For 110 volt, 25 cycle A. C. in **$24.50** cabinet
Cat. No. 1388

25 cycle chassis and speaker **$22.00** only
Cat. No. 1389

For 220 volt, 60 cycle A. C. in **$23.50** cabinet
Cat. No. 1390

220 volt, 60 cycle chassis and **$21.00** speaker only
Cat. No. 1391

For 110 or 220 volt D. C., using three 236's, one 237, two 238's **$23.50** (Push-Pull Pentodes), in cabinet Cat. No. 1392

D. C. chassis and speaker only...... **$21.00**
Cat. No. 1393

YOU will agree that our decision in selecting quality at a low price is in keeping with our motto: "The customer must be satisfied."

Our new line of radio and phonograph-radio combination receivers were thoroughly tested over a long period of time to assure long life. We took into consideration tone, selectivity, durability and appearance. There can be no reason for dissatisfaction with the extra careful design of these receivers.

Constant, non-erratic performance is assured, as all circuits are carefully shielded and wired to prevent any parasitic coupling of radio and audio frequency currents. The power transformers, chokes and condensers are amply large to insure constant humless operation. Tubes last longer in these receivers, due to the excellent voltage control of oversized transformer.

All receivers are quoted less tubes

American
6 Tube All-Wave
Midget Receiver
Variable Mu, Push-Pull Pentode

Covers 200 - 2000 Meter Band.

Uses two 235, one 224, two 247, one 280 tubes

No plug-in Coils. Incorporates change over switch for changing wavelength. 3 gang Condenser (3 tuned circuits); Tone Control; Volume Control. Full vision dial with traveling Pilot Light and Indicator.

Phono attachment with "Phono-Radio" switch. Rola Dynamic Speaker, 8". Beautiful Walnut Cabinet. Size: 14½ x 19 x 10½". Sensitive, selective, quiet operation with ample volume. This receiver will tune to the regular broadcast ship, commercial, aircraft and foreign broadcast wavelengths.

For 110 volt, 60 cycle A. C. **$31.50** operation
Cat. No. 1361

Above Chassis and Speaker **$29.00** Only.
Cat. No. 1362

For 110 volt, 25 cycle A. C. in **$34.50** cabinet
Cat. No. 1363

25 cycle Chassis and Speaker **$31.75** Only.
Cat. No. 1364

For 220 volt, 60 cycle A. C. in **$33.50** cabinet
Cat. No. 1365

220 volt, 60 cycle Chassis and **$31.00** Speaker Only.
Cat. No. 1366

For 110 or 220 volt D. C. using three 236's, one 237, two 238's **$31.50** (Pentode Push-Pull). In cabinet Cat. No. 1367

D. C. Chassis and Speaker Only. **$29.00**
Cat. No. 1368

At an additional cost of $3.75 the 6 tube American Receivers can be furnished to include the 75-200 meter short-wave band in addition to the broadcast band. A non-inductive selector switch eliminates the use of plug-in coils.

American
7 Tube Super Heterodyne
Midget Receiver
Variable Mu, Push-Pull Pentode

Uses two 235, two 224, two 247, one 280 tubes

3 gang condensers (3 tuned circuits), tone control, volume control, phono attachment with Full vision dial with "Phono-Radio" switch, traveling pilot light. Rola 8" Dynamic Speaker. Beautiful Walnut cabinet Size 19 x 15 x 10½. Hairline tuning, super-sensitive using the Super-Heterodyne circuit. Plenty of volume. Quiet in operation. This receiver operates as well as any good 8-tube Super-Heterodyne.

For 110 volt, 60 cycle A. C. **$27.50**
Cat. No. 1355

Chassis and Speaker only **$25.00**
Cat. No. 1356

For 110 volt, 25 cycle A. C. **$30.50**
Cat. No. 1357

Chassis and Speaker only **$28.50**
Cat. No. 1358

For 220 volt, 60 cycle A. C. **$29.50**
Cat. No. 1359

Chassis and Speaker only **$27.00**
Cat. No. 1360

American
6 Tube 2 Volt Air-Cell
BATTERY MIDGET RECEIVER
Screen-Grid and Pentode Tubes

Uses Four 232's, one 230, one 233 Tubes

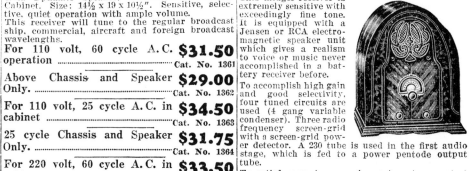

This midget receiver is extremely sensitive with exceedingly fine tone. It is equipped with a Jensen or RCA electromagnetic speaker unit which gives a realism to voice or music never accomplished in a battery receiver before.

To accomplish high gain and good selectivity, four tuned circuits are used (4 gang variable condenser). Three radio frequency screen-grid with a screen-grid power detector. A 230 tube is used in the first audio stage, which is fed to a power pentode output tube.

To satisfy your tone requirements, a tone control has been included. This receiver is also equipped with phonograph pickup connections, with "Phono-Radio" switch. For ease of tuning a full vision dial is incorporated, with illuminated traveling indicator.

Requires an Eveready air cell "A" battery, which should last a year with moderate use; three 45 volts of "B" batteries, and three 4½ volts "C" batteries, which are not supplied at prices quoted.

In Midget Cabinet **$26.50**
Cat. No. 3718

Chassis and Speaker only **$21.50**
Cat. No. 3719

American Sales Company, 44 W. 18th Street, New York, N. Y.

American
Phonograph and Radio
with Dual 78 and 33⅓ R. P. M. Phonograph Motor

6 Tube, Variable Mu, Push-Pull Pentode Receiver with new Polycron Audak Transcription Pick-up

Uses two 235, one 224, two 247, one 280 tubes.

Here is an excellent seller. Not only is it a fine Radio Receiver, but by means of a throw over switch you can operate the phonograph.
Will play all types of records, including the new Victor Long Playing "Transcription" Recordings.
Has a new type Green Flyer Induction Motor which has a dual speed of 33 1/3 or 78 R. P. M., and may be operated at either speed.

An Audak Pick-Up, with volume control, 12" turntable, and automatic stop is incorporated. Rola 8" Dynamic Speaker. Beautiful Walnut Cabinet. Size: 23½ x 17½ x 14". 3 gang Condenser (3 tuned circuits); Tone Control; Volume Control; Full vision dial with traveling Pilot Light and indicator.
Sensitive, quiet, selective and with ample volume.

For 110 volt, 60 cycle A. C.	**$59.50**	Cat. No. 1375
For 110 volt, 25 cycle, A. C.	**$62.50**	Cat. No. 1376
For 220 volt, 60 cycle A. C.	**$61.50**	Cat. No. 1377

Same Combination with a 7-tube Super-Heterodyne Chassis
(Described on other page)

For 110 volt, 60 cycle A. C.	**$63.50**	Cat. No. 1378
For 110 volt, 25 cycle A. C.	**$66.50**	Cat. No. 1379
For 220 volt, 60 cycle A. C.	**$65.50**	Cat. No. 1380
For 110 volt D. C., using three 236's, one 237, two 238's (Push-Pull Pentodes)	**$59.50**	Cat. No. 1381

American
Grandfather Clock Radio

6 Tube Variable Mu, Push-Pull Pentode
Uses two 235, one 224, two 247, one 280 tubes

Has a Sessions Electric, Noiseless Clock; Rola 8" Dynamic Speaker; Caswell-Runyon Cabinet of Matched Oriental Striped Walnut. Size 68 x 15 x 11".
3 gang Condensers (3 tuned circuits); Tone Control, Volume Control, Phono Attachment, with "Phono-Radio" switch. Full vision dial with traveling Pilot Light and Indicator.
Selective, sensitive, humless operation with ample volume.

For 110 volt, 60 cycle A. C.
$43.50
Cat. No. 1382

For 110 volt, 25 cycle A. C.
$46.50
Cat. No. 1383

For 220 volt, 60 cycle A. C.
$45.50
Cat. No. 1384

For 110 or 220 volts D. C., using three 236, one 237, two 238 tubes (Push-Pull Pentode)
$45.50
Cat. No. 1385

1932 Catalog Page

Grandfather Clock Radio
with 7-tube Super-Het Chassis
(Described on other page)

110 volt, 60 cycle A. C.	**$47.75**	Cat. No. 3705
110 volt, 25 cycle A. C.	**$51.50**	Cat. No. 3706
220 volt, 60 cycle A. C.	**$49.75**	Cat. No. 3707

American
6 Tube Console Receiver
Variable Mu, Push-Pull Pentode

Uses two 235, one 224, two 247, one 280 tubes.

Beautiful Caswell-Runyon Walnut Cabinet, 39" high.

3 gang Condenser (3 tuned circuits).
Full vision dial with traveling Pilot Light. Tone control; volume control and phono attachment with "Phono-Radio" switch.
Rola 8" Dynamic Speaker. Selective, sensitive, quiet operation, with ample volume.

LOW-BOY

For 110 volt, 60 cycle A. C.	**$39.75**	Cat. No. 1369
For 110 volt, 25 cycle A. C.	**$42.75**	Cat. No. 1370
For 220 volt, 60 cycle A. C.	**$41.75**	Cat. No. 1371

Console with 7-tube Super-Het Chassis
(Described on other page)

For 110 volt, 60 cycle A. C.	**$43.25**	Cat. No. 1372
For 110 volt, 25 cycle A. C.	**$46.50**	Cat. No. 1373
For 220 volt, 60 cycle A. C.	**$45.50**	Cat. No. 1374

American
6 Tube 2 Volt Air-Cell
BATTERY CONSOLE RECEIVER
Screen-Grid and Pentode Tubes
Uses four 232, one 230, one 233 tubes

This midget receiver is extremely sensitive with exceedingly fine tone. It is equipped with a Jensen or RCA electro-magnetic speaker unit, which gives a realism to voice or music never accomplished in a battery receiver before.
To accomplish high gain and good selectivity, four tuned circuits are used (4 gang variable condenser). Three radio frequency screen-grid with a screen-grid power detector. A 230 tube is used in the first audio stage, which is fed to a power pentode output tube.
To satisfy your tone requirements, a tone control has been included. This receiver is also equipped with phonograph pickup connections with "Phono Radio" switch. For ease of tuning a full vision dial is incorporated with traveling Pilot light and indicator.
Requires an Eveready air cell "A" battery, which should last a year with moderate use; three 45 volts of "B" batteries and three 4½ volt "C" batteries, which are not supplied at prices quoted.

LOW-BOY

In Console Cabinet	**$42.50**	Cat. No. 3720
Chassis and Speaker	**$21.50**	Cat. No. 3719

Bremer-Tully 6 Tube A. C. Pentode Receiver

Model 80 A. C.

In Lowboy Cabinet

Bremer-Tully is known the world over for its precision laboratories, having built parts for the set builder and the amateur in early days of broadcasting. It is obvious that a manufacturer who could design such quality radio parts should be qualified to design and build an excellent receiver.

Uses four 227 tubes in three stages of ultra-selective Tuned Radio Frequency and a tuned Power Detector (Four gang condenser). A special sensitivity control is embodied on the front panel. A Pentode 247 is used in super-power Audio Amplification with a 280 supplying full-wave rectified current to the tubes.

A Rola Dynamic Speaker, with a special Pentode matching transformer, gives life-like reproduction of radio programs and is powerful enough to reproduce the amount of volume delivered by the Pentode Tube. Properly by-passed with a 25 mfd. condenser for the bias of Pentode tube, giving pure bass reproduction.

The cabinet is substantially built of grained walnut and stands 40 inches high, bringing tuning panel to eye level when sitting. Its fine finish harmonizes it with the surroundings of any home furnishing. Operates on 110-120 Volts, 50-60 Cycle A. C. current.

$29.75
Cat. No. S-1962

19

20

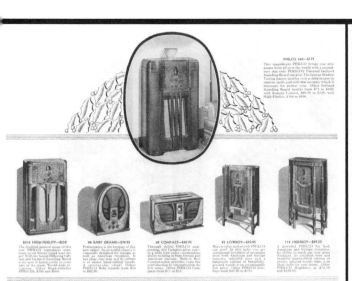

"HOW CAN WARDS DO IT?"

Two Prominent Radio Manufacturers Have Asked Us "How Can Wards Price Their Radios ⅓ to ½ Less Than Other Well Known Makes?"

Here's Wards Answer—

1936 Catalog Page

1. STRAIGHT LINE PRODUCTION
Cuts Radio Manufacturing costs to the bone — just as it has for Automobiles

2. STRAIGHT LINE DISTRIBUTION
From Factory To Wards To You — No Distributor's profits, salesmen's commissions, trade-in allowances or exorbitant advertising costs enter into Wards selling prices

HEAR THE WORLD WITH THESE 7-TUBE A. C. SETS!

Console Model

$32.95
CASH PRICE

or $4 DOWN
ON WARDS BUDGET PLAN

•

162 B 175—Console Model
11¼ by 21¼ by 36 inches high. Ship. wt. 54 lbs. *Not Mailable.* Cash.....**$32.95**
Budget Plan Price: $4 Down, $5 a Month.........**$36.45**

FEATURES for which you'd expect to pay much more — some of which are found only on Wards Airlines.

World Wide Range—Brings in coast to coast Broadcast reception, the best of domestic and foreign short wave stations, as well as amateurs, airplanes, ships at sea, and many police calls.

Instant Dialing—Easier to use than any other dial. Lists about 120 leading Broadcast stations—by call letters. No numbers to remember. Leading short wave channels indicated by numbers on bottom half of dial. Has usual kilocycle scale, too.

Personal Tone Control — The turn of a knob gives you choice of any tone emphasis from deepest bass to brilliant treble.

Automatic Volume Control — Set the volume where you want it—and it remains there without "fading" or "blasting".

Metal Tubes — Used in every socket possible and as additional types become available, will be added as regular equipment.

Beautifully Styled Cabinets — Both Mantel and Console of selected woods. Hand rubbed and hand polished walnut finish.

Gives outstanding performance because it was perfected by countless patents from the laboratories of R. C. A. and Hazeltine. Cushioned chassis and tuning condenser mountings. Wave bands cover from 5.7 to 18 M.C. and 535 to 1730 K.C. Convenient knob on panel switches from one to the other. Large Super-Dynamic Speaker gives clear, natural reproduction and amazing realism of tone.

Even if you hadn't thought of buying a new Radio, be sure to have a demonstration of one of these splendid sets—in your own home—for a full week. The small down payment (or full cash) will bring the radio. If you don't agree it's the best dollar for dollar radio value you ever saw, the finest performing set ever sold at so low a price, return it and we'll refund all your money, even freight charges. But you'll have the surprise of your life when you see so much quality—for so little money. That's why *Wards are the World's Largest Retailers of Radios.*

COMPLETE with Tubes, Wards Standard Quality Aerial and instructions for installation. *For use on 110 to 120 volt, 50 to 60 cycle A.C. only.* 25-cycle models $2.50 extra and shipped separately from factory in Chicago. *State Voltage and Cycles.*

Mantel Model

$26.95
CASH PRICE

or $3 DOWN
ON WARDS BUDGET PLAN

•

462 B 177 — Mantel Model
10¼ by 14¾ by 17 in. high. Ship. wt. 32 lbs. *Mailable.* Cash Price.........**$26.95**
Budget Plan Price:
$3 Down, $5 a Month.**$29.95**

WARDS SETS THE PACE IN RADIO STYLING AS WELL AS IN RADIO VALUES!

WORLD RANGE METAL TUBES

$19.95
CASH PRICE

or $3 DOWN
ON WARDS BUDGET PLAN

•

VERIFIED $30 VALUE!

MAKING RADIO HISTORY!

$14.95
CASH PRICE

BUDGET PLAN
SEE PAGE 689

•

FULL AIRLINE QUALITY

5-Tube A. C. Yes, our shoppers actually found they had to pay $30 to get a Radio equal in quality and performance. Even then it lacked the smart styling of this new Mantel cabinet. Only selected woods are used with a richly grained walnut finish. So compact that it fits nicely on an end table or into a bookcase. This is a set you'll be proud to have in your home.

Long and Short Wave to get the best that's on the air; foreign stations, ships at sea, airplanes, police calls and all your favorite Broadcast programs. The turn of a knob gives you choice of three wave bands.

Large Airplane Type Dial makes tuning easy. Both broadcast and short wave scales shown.

Automatic Volume Control prevents fluctuation. Does away with "fading" and "blasting". **The New Metal Tubes**, radio's latest achievement, used wherever possible . . . more types will be added as they become available.

Latest Type Super-Dynamic Speaker for clear, rich tone and life-like reproduction.

Size 15 by 8¼ by 9½ inches high. Complete with tubes, Wards Standard Quality Aerial and complete instructions. *For 110 to 120 volt, 50 to 60 cycle A.C. only.* 25-cycle model $2.50 extra and shipped separately from factory in Chicago. Ship. wt. 21 lbs. *Mailable.*

462 B 235—Cash Price.................**$19.95**
Budget Plan Price: $3 Down, $4 a Month. **22.45**

Imagine—Famous Airline Quality 5-Tube Receiver for only $14.95! Sets selling for $5 more can't match it for looks or for performance!

Coast to Coast Broadcast Reception.

Full Vision Airplane Type Dial makes station finding easy and pleasant.

Automatic Volume Control reduces annoying "fading" to a minimum.

Powerful Super Heterodyne Circuit, like all Airline Radios, has been perfected with patents from *RCA* and *Hazeltine Laboratories.*

Nowhere have we seen a low priced Radio with the "Class" appearance of this one. Styled in the smart, simple lines of today's design trend. Lovely walnut finish. Take this or any other Airline into your home for a week's trial. Every cent refunded, even freight charges, if you're not completely satisfied. Size 14¼ by 8¼ inches high. Complete with tubes, Wards Standard Quality Aerial, instructions. *For 110 to 120 volts, 50 to 60 cycle A. C. only.* 25-cycle model $2.50 extra and shipped separately from Chicago. Ship. wt. 20 lbs. *Mailable.*

462 B 233 Mantel Radio Complete.....**$14.95**

MONTGOMERY WARD CS **367**

23

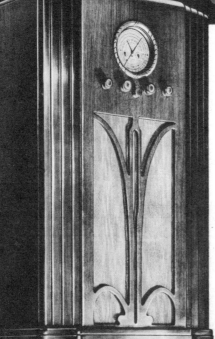

Emerson Radio
1936 Performance

MODEL 111 COMPACT

6 Tube AC-DC Superheterodyne

- Incorporating METAL TUBES
- Emerson Micro-Selector Tuning
- 3 Bands—American, Foreign Short Wave, Regular Broadcast
- Convenient Band Selector Knob
- Dust-proof Dynamic Speaker

Notwithstanding its small size, the Model 111 is a veritable giant for bringing in stations from all corners of the globe. Three Wave Bands, separately controlled by convenient band selector knob on front of cabinet. 19 to 55 meters, covers both day time and night time Foreign and American Short Wave Bands, 63 to 190 meters include American short wave range, both Police Bands, Amateur and Airplane Stations. 180 to 555 meters is the regular broadcast band.

Among its many outstanding engineering features are the telegraphic interference trap, the capacitive filter, the shock mounted variable condenser, and Cadalyte plated chassis. Six tubes, three of which are of the new Metal Types, are arranged in a superheterodyne circuit of extraordinary range. Automatic volume control prevents fading and keeps volume at an even level. Tone control —continuously variable, allows adjustment of tone to any degree as desired by the listener.

Indirectly illuminated airplane dial with etched gold face and Emerson Micro-Selector Tuning principle give the utmost in selectivity and sensitivity. The new color-matched wave band indication is used—red, green and blue flood lighted on dial corresponding to same color as switch knob. Full sized, dust-proof dynamic speaker. Operates on any AC or DC current—110 to 120 volts—25 to 60 cycles.

The gracefully designed cabinet is made of walnut throughout, front and back. Front panel of Burl Walnut. Molded base of solid walnut. Hand-rubbed finish. Size: 10½" high, 13" wide, 6¾" deep. Uses the following tubes: 1-75, 1-43, 1-25Z5, 1-6A8 (metal), 2-6K7 (metal). Listed by Underwriters' Laboratories. Shipping weight, 17 lbs.

No. 9H4206. Emerson Model 111, 6 Tube AC-DC Superheterodyne, complete with RCA Radiotron tubes. List $44.95. Dealer's price, each, $26.97, less 2%, net..

26 43

MODEL 117

5 Tube AC Superheterodyne

- Emerson Micro-Selector Tuning
- Indirectly Illuminated Airplane Dial
- Two Bands—Short Wave and Regular Broadcast—also gets Police, Airplane and Amateur Calls
- Automatic Volume Control and Tone Control
- Dustproof Dynamic Speaker

A 5 tube AC Superheterodyne with two wave bands that has all the beauty and quality for which the Emerson line is famed. That's the Model 117 in a nutshell. 40 to 136 meters cover complete American and Foreign Short Wave Range, including both police bands, amateur and airplane stations. 172 to 555 meters is the regular broadcast band. Tuning is simplified by an indirectly illuminated airplane type dial which incorporates the Emerson Micro-Selector tuning principle. Convenient band selector knob on front of cabinet.

While five tubes are used, seven tube performance is achieved through the use of two multiple-purpose tubes. A Power Pentode Audio Circuit attains high undistorted output of three watts. Automatic volume control prevents fading while tone control allows adjustment of bass or treble response as desired. Other features include shock-mounted variable condenser, mounted on pure gum rubber; telegraphic interference trap; capacitive filter and transformer; and dustproof Cadalyte plated chassis. Large dynamic speaker.

An upright table model of striped American walnut with sides and front made of one continuous piece of curved walnut deflects the modern trend in cabinet design. Front is trimmed with stripes of contrasting inlay. Top designed in raised effect with rounded corners. Pedestal base of triple beaded walnut. Walnut knobs, streamlined to match style of cabinet. Rich walnut finish. Size 15¾" high, 12¼" wide, 8½" deep. Uses the following glass tubes: 1—6A7, 1—6D6, 1—85, 1—42, 1—80. Shipping weight, 21½ lbs.

No. 9H4219. Model 117, 5 Tube AC Superheterodyne, complete with 5 R.C.A. Radiotron tubes. List, $29.95. Dealer's, each, $20.40. Lots of 3 each, $19.75, less 2%, net..................

19 35

Sentinel 2-Volt BATTERY RADIOS

5-TUBE SUPERHETERODYNE

Airplane Dial with Visual On and Off Indicator ● 175 to 550 Meters Gets Police Calls ● Automatic Volume Control ● Extremely Low Battery Drain.

Employs an efficient superheterodyne circuit which develops extreme sensitivity and selectivity for reception of distant stations, which it brings in with abundant volume. A large airplane dial simplifies tuning and a visual on and off indicator prevents leaving the set turned on by mistake. Full automatic volume control prevents fading and distortion. Economical to operate, drawing only ½ Amp. from A battery and 14 milliamperes from B battery. Equipped with a 6" Magnetic speaker. Tubes used are as follows: 1-C6, 1-34, 1-B5, 1-33, 1-5-E1.

Cabinet of selected veneers, has a distinctive maple overlay on the top front of the panel with a sliced walnut center panel and striped walnut front panels overlay sides. There is sufficient room in cabinet for all batteries. Dimensions: 17" high, 14½" wide, 12" deep. Weight, 19 lbs.

No. 9H4230. Sentinel 5-tube, 2-volt battery set, complete with RCA Radiotron tubes, less batteries. List $29.95. Dlr's., ea. $16.55. Lots of 3, ea. $16.00, less 2%, net.............. **15⁶⁸**

All Sentinel 2-Volt Receivers Will Operate On Eveready Air Cell, 2-Volt Dry A Battery or 2-Volt Wet Storage Battery

6-TUBE DUAL-WAVE SUPERHETERODYNE

● **American and Foreign Short Waves** ● **On and Off Visual Indicator in Dials** ● **Automatic Volume Control** ● **Tone Control** ● **Class "B" Amplification** ●

Two wave bands—from 47.5 to 130 meters, and 175 to 550 meters covers many interesting broadcasts, such as foreign short waves, amateur, airplane and police calls, as well as regular broadcasts. Features latest type airplane dial, with visual on and off indicator, automatic volume control, variable tone control and highest selectivity and sensitivity. Class B amplification. Extremely low A and B Battery drain. Equipped with 6" Magnetic speaker. Tubes used are as follows: 1-1C6, 1-34, 1-1B5, 1-30, 1-19, 1-30, 1-6-1. Cabinet of semi-modernistic design, stump and diamond matched walnut veneers, with pin-striped walnut overlays. Hand rubbed finish. Sufficiently large to accommodate all required batteries. Dimensions, 21" high, 15½" wide, 13" deep. Weight, 26 lbs.

No. 9H4231. Sentinel 6-tube 2-volt Table Model, complete with RCA Radiotron tubes less batteries. List $39.95. Dlr's., ea. $22.00. Lots of 3, each, $21.20, less 2%, net......... **20⁷⁸**

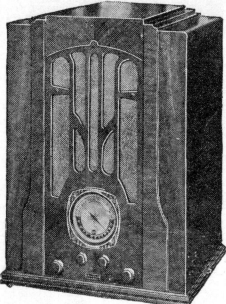

7-TUBE DUAL-WAVE SUPERHETERODYNE

● **3 Bands—16 to 555 Meters ● On and Off Visual Indicator ● Automatic Volume Control ● Variable Tone Control ● Class "B" Amplification ●**

A DeLuxe table model battery receiver, featuring three continuous wave bands from 16 to 555 meters—covering foreign and American short wave, airplane, amateur and police calls, as well as regular broadcasts. A large airplane dial with visual on and off indicator makes for easy tuning. 7-tube superheterodyne circuit incorporates such outstanding developments as three-gang condenser, class B amplification, Full Automatic Volume Control, Variable tone control and many others. Reaches a degree of selectivity and sensitivity rarely ever attained in battery receivers. Equipped with a 6" magnetic speaker. Tubes used are as follows: 1-1C6, 2-34, 2-30, 1-19, 1-5H1. Employs the same cabinet as used with the 6-tube model. Weight, 27 lbs.

No. 9H4233. Sentinel 7-tube 2-volt All-Wave Table Model, complete with RCA Radiotron tubes, less batteries. List, $49.95. Dlr's., ea., $27.30. Lots of 3, ea., $26.50, less 2% **25⁹⁷** Net

CONSOLE MODELS

6-TUBE MODEL

Beautifully designed console with butt walnut front panel and streamlined top construction. Dimensions: 37" high, 22" wide, 12" deep. Equipped with 8" speaker. Weight, 45 lbs.

No. 9H4232. Sentinel 6-tube 2-volt Console, complete with RCA Radiotrons, less batteries. List $59.95. Dlr's., ea., $31.45. Lots of 3, each, $30.25, less 2%, net..................... **29⁶⁴**

7-TUBE CONSOLE

Designed in the modern classic manner, with butt walnut instrument panels, striped walnut side front panels, top and overlay. Dimensions: 39" high, 24" wide, 13" deep. Weight, 57 lbs. Equipped with 8" speaker.

No. 9H4234. Sentinel 7-tube, 2-volt Console, complete with RCA Radiotron tubes, less batteries. List, $69.95. Dlr's., ea., $36.15. Lots of 3, ea., $35.00, less 2%, net..................... **34³⁰**

Refer to Index for Listing on Battery Kits for Sentinel 2-Volt Radios

6-TUBE CONSOLE

7-TUBE CONSOLE

Sentinel AC-DC
MIDGET RECEIVER

- ● Gets Police Calls
- ● Illuminated Tuning Dial
- ● Electro Dynamic Speaker
- ● Licensed Under R. C. A. and Hazeltine Patents
- ● Operates on 110 Volt, 25-60 Cycle, A.C.-D.C.
- ● 4 New Type R. C. A. Licensed Tubes

9 80 NET

TS OF 12

Once More, We Set the Pace!

Chalk up another hit for our buying department—and if you know your radio market you will realize how great a hit this really is. This 4-tube A.C.-D.C. midget is in every sense of the word a precision built job—one that will amaze you with its tonal beauty and true-to-life reproduction.

All the refinements of modern day radio engineering have gone into the design of this little Sentinel model—2-gang condenser—a newer, better audio system—dual purpose tube. These are but a few of the reasons why the Sentinel has achieved such quality of performance—performance which you would ordinarily expect from a 6-tube, or larger, receiver.

The compact little cabinet is attractively designed and sturdily constructed of selected gumwood. Front panel of butt walnut. Beautifully finished. Illuminated full vision dial highlights the beauty of the cabinet and makes for easier tuning. Equipped with built-in aerial and the following new type R.C.A. licensed tubes: 1—12Z3; 1—43; 1—6D6; 1—6C6. Overall size, 10¾x7x 5½. Wgt. 9 lbs.

No. 9H4159. 4 Tube AC-DC. List $17.50. Dlr's., ea. $11.00. Lots of 3, ea. $10.75. Lots of 12, ea. $10.00, less 2%, net...................

9 80 Complete With Tubes

Sentinel 7-Tube ALL WAVE Superheterodyne

- ● Three Bands—16 to 555 Meters
- ● Illuminated Airplane Dial
- ● Automatic Volume Control—Variable Tone Control
- ● Latest Type RCA Radiotron RCA Dual Purpose Tubes

To this handsome 7 tube All Wave receiver go the value honors of the year. Certainly no other radio can offer more for the money. All Wave tuning permits a wide range of reception including Foreign Short Wave, Airplane, Police and Amateur signals, as well as standard broadcasts. All those quality features which go to give a set the utmost in sensitivity and selectivity, are incorporated in this latest type superheterodyne circuit. Among these are, Push-pull amplification, 3-gang tuning condenser, variable tone control and automatic volume control. Employs an electro Dynamic Speaker of proven quality. Illuminated airplane dial together with band control switch simplifies tuning on all wave bands. Has an undistorted output of 4.5 watts. The handsome cabinet, of simple modern design has a genuine walnut front of selected veneers, the front section of panel being sliced walnut with pin stripe overlay side panels. Dimensions: 17″ high, 13″ wide, 9½″ deep. Weight 22 lbs. Tubes used are as follows: 1-6A7, 1-6D6, 1-75, 1-76, 2-41, 1-80.

No. 9H4241. Sentinel 7 tube All Wave Superheterodyne, complete with RCA Radiotron tubes. List $49.95. Dealer's each $29.97. Lots of 3 ea., $29.00, less 2%, net ....

28 42

Sentinel 6-Tube ALL WAVE Superheterodyne

- ● Three Bands—16-555 Meters ● Illuminated Airplane Dial
- ● Automatic Volume Control ● Latest Type RCA Radiotron Dual Purpose Tubes
- ● Variable Tone Control

Give this set the promotion it deserves and we guarantee you a radio business far in excess of what you have ever done before. Features three band tuning, covering Foreign Short Wave, Airplane, Police and Amateur signals, as well as standard broadcasts. Superheterodyne circuit employing 6 of the latest Dual Purpose type tubes, enjoys constructional features rarely found in a set of this type. Automatic volume control and variable tone control, Electric Dynamic Speaker, 3 Watt output. Illuminated Airplane dial calibrates all wave bands in kilocycles. Convenient band selector switch. Cabinet is made of selected walnut veneers throughout. The center of the front panel, of striped walnut, with side panels of striped walnut overlay. Carpathian walnut overlay on top. Measures 20″ high, 15″ wide, 11″ deep. Weight, 20 lbs. Tubes used are as follows: 1-6A7, 1-6D6, 1-75, 2-42, 1-80.

No. 9H4240. Sentinel 6 tube All Wave Superheterodyne, complete with RCA Radiotron tubes. List $39.50. Dealer's, each $21.25. Lots of 3, each $20.00, less 2%, net.......

19 60

1936 Styling Emerson Radio

MODEL 106 Duo-Tone

6 Tube AC-DC Superheterodyne

- Incorporating Metal Tubes
- Emerson Micro-Selector Tuning
  - 2 Bands—American Short Wave and Regular Broadcast
  - Identical Front and Back Design
  - Illuminated Gold Sliding Scale Dial
  - Dustproof Dynamic Speaker

Far in advance of anything competition has to offer, the Model 106 has already shown definite indications of tremendous popularity. Performance of a quality rarely achieved in a small set comes as the result of a superheterodyne circuit, using 6 tubes and taking full advantage of 456 kilocycles as an intermediate frequency. Two of the 6 tubes are of the new metal type, resulting in increased efficiency. Equipped with the telegraphic interference trap and the capacitive filter, the model 106 is remarkably free from man-made static and code signals. Automatic volume control prevents fading and keeps volume desired at an even level.

There are two distinct wave bands—one from 70 to 203 meters, covering the complete American Short Wave range, including both police bands, amateur and airplane stations—the other, from 195 to 565 meters, covering the regular broadcast band. The Emerson Micro-Selector principle of tuning is employed here in an ingenious type of illuminated sliding scale dial. A convenient band selector knob on front of cabinet, permits switching from one band to the other. The full size dynamic speaker is fully impregnated against dust. Cadalyte plated chassis is rust proof. Equipped with built-in antenna.

The uniquely designed streamlined cabinet is of high grade matched American Butt Walnut with Mahogany inlays and ebony base. Back is designed and finished same as front. Hand-rubbed finish. Metalized knobs to match escutcheon. Size 8¾" high, 12¼" wide, 5¾" deep. Operates on AC or DC current—110 to 120 volt—25-60 cycles. Listed by Underwriters' Laboratories. Complete with the following tubes: 1-6A7, 1-6D6, 1-43, 1-25Z5, 1-6H6 (metal), 1-6F5 (metal). Ship. weight, 15 lbs.
No. 9H4205. Emerson Model 106 Duo-Tone, 6 tube AC-DC Superheterodyne, complete with RCA Radiotron tubes. List, $39.95. Dealer's price, each, $18.70. Lots of 3, each, $17.25, less 2%, net...............

16⁹⁰

16.90

MODEL 107 Duo-Tone

6 Tube AC-DC Superheterodyne

- Incorporating METAL TUBES
- Emerson Micro-Selector Tuning
  - 3 Bands—American, Foreign Short Wave, Regular Broadcast
  - Identical Front and Back Design
  - Rust-proof Dynamic Speaker

The Model 107 has the same effect on the eye as on the ear—simply breath taking! Three out of the 6 tubes used in the ingenious Superheterodyne circuit are of the new Metal type and are largely responsible for the excellent performance of which this receiver is capable. Three separate wave bands—19 to 55 meters cover both the daytime and nighttime Foreign and American Short Wave Bands; 63 to 190 meters for complete American Short Wave Range, both Police Bands, Amateur and Airplane stations; 180 to 555 meters is the regular broadcast band.

Equipped with the telegraphic interference trap and the capacitive filter, static and other disturbances are minimized. Automatic volume control and tone control allows the user a wide adjustment of the tone and volume at any desired point. Chassis is Cadalyte plated, rendering it rust proof and durable.

An illuminated dial, employing the Emerson Micro-Selector Tuning principle permits maximum tuning ease.

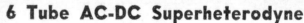

The new color-matched wave band indication is used—red, green and blue flood lighted on dial corresponding to same color on switch knob. Convenient band selector knob on front of cabinet. Full sized dust-proof dynamic speaker.
The cabinet, designed and finished the same in back as in front, is indeed a work of art. Made of high grade matched American Butt Walnut with Mahogany inlay and ebony base. Metalized knobs to match escutcheon. Hand-rubbed finish. Size 10¼" high, 15" wide, 7" deep. Operates on any AC or DC current—110 to 120 volts—25 to 60 cycles. Listed by Underwriters' Laboratories. Uses the following tubes: 1-75, 1-43, 1-25Z5, 1-6A8 (metal), 2-6A7 (metal). Shipping weight, 18 lbs.

No. 9H4207. Emerson Model 107 Duo-Tone, 6 tube AC-DC Superheterodyne, complete with RCA Radiotron tubes. List, $49.95. Dealer's price, each, $29.97, less 2%, net...............

29³⁷

29.37

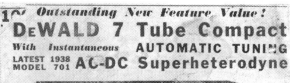

1937 Catalog Page

30

PICKWICK CONSOLE and AUTO RADIO

PICKWICK 5-TUBE AC 2-BAND

The Pickwick Five AC Super, superbly engineered with eight tuned circuits, develops a state of sensitivity uncommon in compact receivers. Two-band tuning: 75 to 175 meters for amateur, short wave and police calls; 175 to 550 meters for Domestic Broadcasts. Large 3¾ inch Gold Dial in 3 colors, greatly enhances the beauty of the receiver and makes for easy tuning. Distant reception without fading assured by perfected, fully automatic volume control. Equipped with full-sized, powerful dynamic speaker. Housed in hand-rubbed walnut cabinet of sloping, one-piece front and top: measures 11" x 7" x 6". Operates on 110-volt, 50-60 cycle, A.C. current only. Tubes: 6A7, 6D6, 75, 41, 80. Shipping weight, 12½ lbs.

No. 723H58—Retail, $22.25. Dealer, each, price **$13.35**

5-TUBE, 2-BAND AC-DC

Same as above, but 5-tube T.R.F., that operates on any 110-Volt current, A.C. or D.C. Tubes: 6D6, 6C6, 43, 25Z5, L49C. Shipping weight, 10½ lbs.

No. 723H59—Retail, $17.00. Dealer, each, price **$10.20**

PICKWICK FIFTEEN-TUBE CONSOLE

MAGIC EYE 3-BAND SUPER

1937 Catalog Page

The RADIO of radios. 15 tubes, each one responsible for practically every individual advance, and scientific development in present-day radio engineering; and every one of these tubes supremely utilized in this receiver to create the masterpiece of modern radio genius. The gigantic four-colored, 8-inch Vernier Spinner Dial, calibrated in one-piece front and top. Cabinet measures 39 x 23 x 11½. Shipping weight, 76 lbs. the dial to the other with a slight touch of the finger-tips, makes Foreign Short Wave and Domestic tuning a delight. AND the Magic Eye assures almost automatic hair-line tuning. AND highly developed coils, with separate coils on each band, make Foreign Reception a reality, not a bug-a-boo.

- 13½ Watt Output
- 12-Inch Concert Speaker
- Giant 8-Inch Gold Dial
- Foreign Reception
- Dual Audio Channel
- Sensitivity Switch
- Bass Compensator
- Tone Control—A.V.C.
- —Magic Eye

The tremendous 13.5-Watt output of the set is handled by a 12-inch concert type dynamic speaker, which, coupled with a bass compensated volume control circuit and dual audio channel, produces auditorium volume with symphonic tones. Continuous variable tone control emphasizes the high and low notes at your pleasure. Three-range sensitivity control permits the extreme sensitivity of the set to be materially reduced for local and short distance reception. The console is a masterpiece of cabinet artistry, with high grade matched walnut, hand-rubbed to a piano finish, a perfect housing for so fine a set. Cabinet measures—Height, 41 ins.; width, 24¼ ins.; depth, 13 ins. Tubes are 3-6K7, 1-6A8, 4-6C5, 2-6H6, 2-6F6, 2-5Y3, 1-6G5.

No. 723H60—Retail, $100.00. Dealer, each, price **$60.45**

SHIPPING WEIGHT 85 POUNDS

PICKWICK TWELVE-TUBE CONSOLE

3-BAND SUPER

The Pickwick TWELVE is basically a superfine radio, engineered for quality in tone, sensitivity and selectivity—and designed for beauty and efficiency. Auditorium volume, combined with a high state of sensitivity on all three bands, make for a valuable musical and electrical instrument. The INTEGRAL UNIT coil system assures Foreign Reception from 17 to 52 meters; Amateur, Foreign, Airplane, Ships at Sea and Police Calls from 52 to 175 meters; Domestic Broadcasts from 175 to 555 meters.

MAGIC EYE

A Giant 8" Dial in four colors, with separate colors for each band (calibrated in meters and kilocycles), is used with the Magic Eye for delightfully easy and almost automatic tuning—a feature that will be a source of pleasure and lend a sense of luxury to the instrument.

12" CONCERT SPEAKER

The twelve-inch Concert Dynamic Speaker handles the auditorium output of the receiver, with a beauty possible only with a combination of bass compensation and tone control. Every aid to perfect audio amplification is employed, that this instrument may deliver bass or treble in an equally thrilling manner.

- Foreign Reception
- 8-Inch Gold Dial
- Bass Compensator

The Console is a masterpiece of cabinet artistry, made of high grade butt walnut, hand-rubbed to a piano finish, trimmed with horizontal pilasters on each side, and a round, one-piece front and top. Cabinet measures 39 x 23 x 11½. Shipping weight, 76 lbs. Tubes: 2-6K7, 2-6H6, 2-6F6, 2-5Y3, 2-6C5, 1-6G5, 1-6A8.

No. 723H61—Retail, $77.00. Dealer, each, price **$45.10**

PICKWICK 6-TUBE SPECIAL AUTO RADIO

The Pickwick 6-tube Auto Radio, in its gray case, trimmed in chromium, is a car radio of beauty; its twelve tuned circuit superheterodyne is an engineering feat. Equipped with tone control, automatic volume control, long and short distance sensitivity switch, full-sized, powerful, 6½" dynamic speaker, and plug-in provision for extra speaker, the Pickwick has every refinement of the ultra-modern Auto Radio. Tuning controls for steering post, under-dash and custom-dash mountings, to match all cars so provided, included in price. Dimensions: 9½" x 6¾" x 8". Tubes: 2-6D6, 6A7, 6Q7, 42, 0Z4. Shipping weight, 22 lbs.

No. 723H62 — Including controls. Retail, $38.00. Dealer, each, price **$22.85**

CUSTOM BUILT DASH MTGS.

Tuning controls custom-built, to match the dash of all modern automobiles, included in quoted price. Where no provision has been made for dash mounting, steering post or under-dash mounting available. When ordering state make and model of car.

PICKWICK 7-TUBE DE LUXE AUTO RADIO

The Pickwick 7-tube Auto Radio is handsome in its chromium and red trimmings on its crackled gray case (illustrated). Its superbly engineered superheterodyne circuit has a sensitivity so far in excess of ordinary requirements, that a sensitivity switch is included in its equipment to cut down the extreme sensitivity of the set for local reception.

A powerful, full-sized dynamic speaker capably handles the power of the set. Tone Control accommodates the pitch to the acoustic requirements of your car and to your individual tastes. Automatic Volume Control prevents station fading. Provision is made for extra speaker, if desired. Every provision made for ease of installation. Custom-built equipment to match the dashboard of all modern cars included in price.

Multi-purpose tubes are 6K7, 6L7, 6K7, 6Q7, 6C5, 6F6, 0Z4. Dimensions: 10½ x 6¾ x 8. Shipping weight, 22 lbs. No. 723H63—Complete, including controls. Retail, $43.75.

Dealer, each, price **$26.25**

PHILCO AUTOMATIC TUNING

On every Automatic Tuning Philco, there's a dial like that on an automatic telephone. Around are the call letters of all your favorite stations.

But with Philco Automatic Tuning you dial only once! One glance shows the station you want. A flick of your fingers . . . and *CLICK* . . . there it is! Tuned *instantly, silently, perfectly!*

One motion is all you make. Back of the dial, the invisible fingers of Philco Magnetic Tuning complete the job with infallible precision. The program comes in more perfectly tuned than eye or ear could do it . . . and stays perfectly tuned as long as you choose to listen!

When you want American stations that you listen to less frequently . . . you simply tune the ordinary way. Foreign stations are tuned by name. Once you have the station, Philco Magnetic Tuning assures the greatest possible accuracy, the finest possible tune!

$10 DOWN brings you *any* Automatic Tuning PHILCO

Why not begin enjoying all the convenience of Philco Automatic Tuning in your own home? For a down payment of only ten dollars, you can have *any* Automatic Tuning Philco installed and playing in your own living-room. By special arrangement with the Commercial Credit Company, your Philco dealer will install even the Philco 116X DeLuxe for ten dollars down!

In this model Philco Automatic Tuning is combined with Philco High-Fidelity and the Philco Foreign Tuning System! Favorite American Stations are tuned *automatically.*

You enjoy all the realism of Philco High-Fidelity reception with all the tones *and overtones* that make voices and instruments so thrillingly true to life. Acoustic Clarifiers eliminate "boom" from the low notes, while every note is brought up to ear level by the Philco Inclined Sounding Board. Five Spread-Band Tuning Ranges cover all that's interesting in the air . . . with all important overseas stations named, located and spread six times farther apart on the Philco Spread-Band Dial!

See your classified telephone directory for your Philco dealer.

PHILCO AUTOMATIC TUNING MODELS...$100 up • PHILCO TABLE MODELS...$20 up • PHILCO CONSOLES...$39.95 up • PHILCO AUTO RADIOS...$19.95 up • *All prices less Aerial*

1937 Magazine Ad

I've got it, Mommy!

Now . . . LABYRINTH RADIO with FLASH TUNING

You'll be delighted, too, with this Automatic Tuning Stromberg-Carlson. A twist of a knob, and—*flash*—there's the station perfectly tuned, *restored*—with the call letters lighted up. And the knob is the same one you use for regular tuning—that's the best of it—no complications, no bother. Perfected Automatic Frequency Control keeps the selected station accurately tuned.

What a radio to have in your home—a Stromberg-Carlson with the "Labyrinth," most famous invention for giving pure tone; this new Automatic Flash Tuning, and a cabinet of real style!

Prices range from $57.50 to $850. (Slightly higher in Southeastern States and West of the Mississippi.) New edition of booklet, "How to Choose a Radio," *free* at Stromberg-Carlson dealers, listed in the classified section of your telephone directory, or by mailing the coupon at lower right.

A LABYRINTH RADIO only $15.00 down
Other Stromberg-Carlsons as low as $5.00 down

There Is Nothing Finer than a
Stromberg-Carlson

1937 Magazine Ad

Lounge and Listen

PHANTOM BACHELOR—One of five striking Arvin chair-side models

IT'S AN ARVIN • THE NEW BIG NAME IN RADIO

● How beautiful this radio will look beside your favorite easy chair, or at the end of your davenport, within easy reach, as you lounge and listen! Only one of many new Arvin Radios designed as useful and attractive pieces of furniture, as well as fine radios. Other examples—that unusual bookcase design below—a handsome addition to your furnishing scheme. And that trim little table set for your bedroom . . . But wait—inside these handsome cabinets lies the newest improvement in radio reception . . . The Phantom Filter Circuit. An exclusive Arvin development that reduces noise astonishingly. And of course, as noise goes out, better tone quality comes in—and station range is increased . . . Drop in where Arvins are shown and *see* these new ideas in radio. Over 30 models—strikingly styled—acoustically correct in design—all-wave reception—large-face dials—automatic tuning in larger models.

Only Arvin has the Phantom Filter Circuit

MODELS FROM
$19.95 TO $175
Slightly higher in extreme South and west of Denver

THE PHANTOM QUEEN (Left) New! A radio in handsome walnut bookcase design. 12 tubes. Twin F speakers. Automatic tuning with new Arvin Presto-Station-Changer. Flip the dial—there's your station.

THE PHANTOM PRINCE (Right) Popular priced console, beautiful in line and finish. Automatic tuning with new Arvin Presto-Station-Changer. 12 tubes. Many handsome consoles to choose from.

THE PHANTOM BABY—(Center)—Powerful five-tube inexpensive table model. In hand-rubbed walnut or antique white. Typical of Arvin's smart table sets.

NOBLITT-SPARKS INDUSTRIES, INC. • COLUMBUS, Indiana
Also Makers of Arvin Car Heaters and Arvin Car Radios

1937 Magazine Ad

CROSLEY ALL-STAR RADIOS FOR 1938

MAKE THE 5-POINT COMPARISON TEST

SUPER 11 TUBE $69.95

11 TUBES ELECTRIC TUNING $89.95

6 TUBE CHAIRSIDE ELECTRIC TUNING $64.95

SUPER 6 TUBE COMPACT $34.95

DEFINITELY, IN RADIO . . . THE SWING IS TO CROSLEY

THE CROSLEY RADIO CORPORATION - CINCINNATI
Powel Crosley, Jr., President
Home of "the Nation's Station"—WLW—500,000 watts—70 on your dial

PRICES IN WEST AND SOUTH SLIGHTLY HIGHER

FIVER $24.95

FIVER CHAIRSIDE $27.95

SUPER 6 TUBE COMPACT $44.95

FIVER ROAMIO AUTOMOBILE RADIO $19.95

1937 Magazine Ad

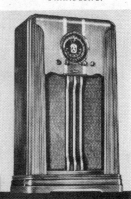

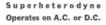

36

M-m-m — just listen—

There Is Nothing Finer than a

Stromberg-Carlson

FEATURES THAT MAKE THE 1938 STROMBERG-CARLSONS OUTSTANDING

Acoustical Labyrinth
The long, winding passageway of the Labyrinth, takes the place of the small box that carries the cabinet which is the secret of the exaggerated bass in low tones. It gives you deeper bass notes, with new fidelity.

Carpinchoe Speaker
The edge suspension of Carpinchoe Leather is more capable of moving absorbing the vibrations at the edge of the cone than any other material, thus safeguarding the famous Stromberg-Carlson tone.

Flash Tuning
Visual Automatic Station Finding in which you see the station you've tuned to up when you touch it. Automatically. Exclusive Control makes Flash Tuning attractive tuning.

Amazing "Labyrinth" Tone Entrances Everyone Who Hears It

Tone you can lose yourself in, so natural is it; reproduction of everything your heart thrills to in radio from a magnificent symphony—to the rich laughter of your favorite radio comedienne, all with "footlight" fidelity.

And beauty—it sounds extravagant to speak of radio cabinets which assail your sense of beauty by the artistic forming and exquisite matching of their rare woods, by the symmetry of their contours and proportions, but the new Stromberg-Carlsons do exactly that.

And—those who have been craving just such fineness revel in them.

Then, there are new and exclusively Stromberg-Carlson details such as Flash Tuning, which lets you see the station you have automatically tuned; Carpinchoe Leather Speaker Suspension; Statuary Bronze Dials with wide band spread.

The price range of Stromberg-Carlsons is from $57.50 to $1050; Antenna Kit $7. (Slightly higher in Southeastern States and West of Mississippi). Booklet, "How To Choose a Radio" may be obtained from authorized dealers listed in your classified directory or by mailing coupon.

Stromberg-Carlson Telephone Mfg. Co.
165 Carlson Road, Rochester, N. Y.
Send illustrated booklet "How to Choose a Radio."

Name ....
Street ....
City .... State ....

1937 Magazine Ad

The Biggest Radio Buys of 1938!

THESE examples chosen from the new 1938 General Electric Radios represent good news for radio buyers. Never in G-E history has so much beauty, performance, and sheer value been offered for your money. See and hear them soon—at your nearest G-E Radio Dealer's. This year, more than ever, you'll get more when you buy a new 1938 G-E.

Your General Electric radio dealer will make a liberal allowance for your present radio and arrange convenient terms of payment.

with the new and exclusive G-E TONE MONITOR
- New cabinet styling—hand rubbed finishes
- New Louver Dials
- Visual Volume and Tone Indicators
- American and Foreign Programs
- Expanded Tone Range
- Police, Amateur and Aircraft Calls
- New Stabilized Dynamic Speakers

MODEL F-70—(at right) 7 tubes, 3 bands—TONE MONITOR, Louver Dial—Foreign and Domestic Stations—Police, Aircraft, Amateur Calls—Large Dynamic Speaker—5 watts output.

MODEL F-53 5 tubes, 2 bands—Edge-lighted Dial—Domestic Stations, Police, Amateur Calls—Large Dynamic Speaker.

WITH THE AMAZING NEW G-E TONE MONITOR—
MODEL F-65—6 tubes, 2 bands—TONE MONITOR, Louver Dial—Domestic and Foreign Stations—Police, Amateur and Aircraft Calls—Large Dynamic Speaker—5 watts output.

12 Super Value G-E Models to select from $19.95 Up

Prices slightly higher South and West—subject to change without notice.

The New GE RADIO — GENERAL ELECTRIC

FOR REPLACEMENTS SPECIFY GENERAL ELECTRIC PRE-TESTED TUBES

1937 Magazine Ad

They're New! Thrilling!

Enjoy these Latest Triumphs of Philco Research in

Glorious Tone and Superb Performance!

YOURS FOR ONLY A FEW PENNIES A DAY!

YEAR by year, the scientific achievements from the great Philco laboratories have made radio tone more gloriously real, performance more thrilling and the joy of fine radio entertainment more easily within the reach of every home. Today, the owners of almost twelve million Philco radios enjoy the fruits of the inventive genius of Philco engineers. Wherever you live, Philco is famous for that finer tone and performance which is born of the millions of dollars spent in scientific research.

Whatever the purpose, whatever the price, your nearest dealer offers a Philco to suit your need. Visit him today. He'll gladly demonstrate how Philco gives you the utmost in radio pleasure for the price you wish to pay.

TRANSITONE... (Left)
Sold and Guaranteed by PHILCO
New Philco inventions, new tubes, new speaker refinements give you amazingly rich tone and powerful performance never before heard in radios of this compact size. And, for the first time, Underwriters Laboratories, Inc. approval! No lost wires, SAFE from fire and shock, SAFE for your home and children! Several models, beginning at only $9.95, full complete price including built-in aerial.

THE TRAVELER... (Right)
"She shall have music wherever she goes!"
An entirely new kind of radio, invented by Philco engineers ... the Philco 7T. Portable ... self-powered ... needs no aerial, ground or "house-current". Take it with you wherever you go ... traveling, on trains, in hotels. Use it in camps, cottages, boats, at bathing beaches—anywhere indoors or outdoors. It plays, without "hooking up" to anything!

PHILCO MYSTERY CONTROL
The great Philco 116RX, above, gives your radio's finest in depth and richness of tone, in superb American and Foreign reception, in beauty of design and costly cabinet woods. In addition, it brings you Mystery Control—the most miraculous radio invention since radio itself. Remote Control, without wires or plug-in connections to radio, electric outlet or anything else. Wherever you are in your home, you change stations, adjust volume and turn the radio off without the annoyance of jumping up and running to and fro. In many cases, your present radio may be traded in as part payment and the easiest of monthly terms arranged for the balance.

THE NEW SPINET FURNITURE DESIGN
The new vogue in radio furniture, acclaimed by home decoration experts everywhere. Exquisite simplicity ... graceful styling that adds to the beauty of your home, that does not clash with its surroundings. Philco Spinet style models give you the finest performance features including Electric Push-Button Tuning on 8 stations. $10 down puts the 40XX (illustrated) in your home!

PHILCO
A Musical Instrument of Quality

LOVELY, STREAMLINE TABLE MODELS
If you prefer fine radio tone and performance in smaller space, Philco table models are your answer! Powerful American and Foreign reception, Electric Push Button Tuning, all the costly features of a big radio. Housed in table cabinets of handsome design in richly finished, inlaid Walnut woods. Just $5.50 down puts the 39T (illustrated) in your home!

PHILCO ALSO OFFERS YOU—
a complete selection of Battery Radios for Farm Homes, Phonographs with Philco, Philco Auto Radios, Philco High-Efficiency Radio Tubes and the new vertical Philco Safety Aerial.

1938 Magazine Ad

"None of my girl friends has a radio as grand as this little Arvin, Mother."

"It's lovely, Jane, and I've asked Dad to get a big Arvin for the living room."

Lovely to look at ... and lovely to hear! This smartly styled little Arvin Rhythm Baby, shown in Jane's room, is the answer to the prayers of thousands of girls and boys for personal radio service. This smallest and lowest priced Arvin will thrill anyone with its fine performance and delightful beauty. It's a perfect "personal" set to supplement the family radio in your living room—where you should have one of the splendid Arvin Rhythm Consoles or larger table models. There's an Arvin Rhythm Radio to fit every purse and purpose—and every model is a beautiful piece of furniture as well as an inspiration for increased family happiness.

The Arvin Rhythm Baby, Model 417, shown in Jane's room, is priced at only $15.95. Arvin Rhythm Baby, Model 402, above, in walnut, more or marbleized cabinet at your choice. $24.95. Arvin Rhythm Queen, Model 527, at only $19.95 offering controls in matched green walnut cabinet. $18.95. Other Arvin consoles as low as $49.95 and up to $175. Slightly higher Denver and West.

GIVE YOUR YOUNGSTER A *Personal* Arvin
Treat the Family to an Arvin Downstairs, Too!

ARVIN RHYTHM RADIOS

With such a variety of splendid programs on the air, everyone should have the privilege of hearing the entertainment of his choice. Father and Jack want the ball game. Mother enjoys the symphony. Sister likes to listen to her favorite dance band. And a foreign-station-fan in every family is not unusual. Obviously, no single radio can satisfy everyone at the same time.

The new 1937 Arvin Rhythm Radios were designed and built to satisfy the desires of everyone in your family—and the reasonable prices enable you to buy the two or more Arvins you need for complete radio enjoyment in your home. The little "personal" Arvins are excellent radios. The bigger Arvins bring world-wide reception to those who want it—and the way they reproduce American broadcasts is beyond description. If you live out beyond the power lines—or spend week-ends in a country cabin—there's an Arvin battery radio that will fit your needs perfectly.

The new Arvins have every modern feature known to radio science—and many exclusive Arvin developments, such as the traveling spotlight station finder on all-wave models—big reverse lighted clock-face dials—high-low-speed tuning drive controlled by one knob. See and hear the new Arvins—sold on easy terms by most dealers.

NOBLITT-SPARKS INDUSTRIES, Inc., Columbus, Ind.
Also makers of Arvin Hot Water Car Heaters

ENJOY THE SAME FINE RADIO RECEPTION IN YOUR CAR, TOO
Now, in your car, you can have a radio that gives you the same high quality reception that Arvin Rhythm Radios bring to your home. Arvin Tailor-Fit Car Radios are distinguished by their clean distinctive appearance—with overhead, in-the-set, or dash type speakers—instrument panel or steering column control. Priced as low as $39.95.

ARVIN *Rhythm* RADIOS

1939 Magazine Ad

38

1938 Catalog Page

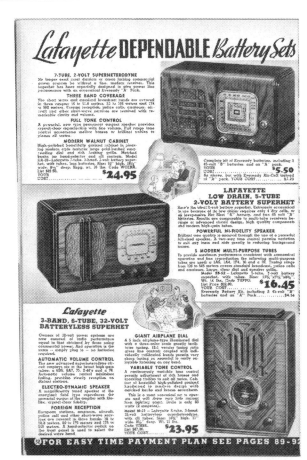

1938 Catalog Page

1938 Catalog Page

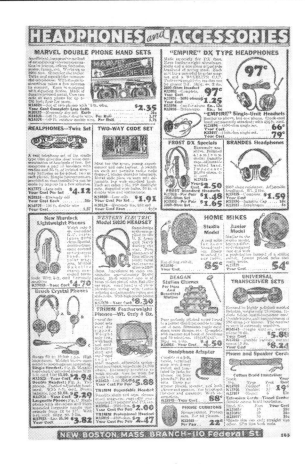

1938 Catalog Page

39

8 Tube 3 band SUPERHET
TOUCH A BUTTON---that's all

De Luxe Instruments

Lafayette

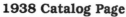

TABLE MODEL B25

$58.50

$49.50

1938 Catalog Page

PHONO-RADIO *Combination*

ELECTRIC
Touch Tuning

WORLD WIDE SHORT-WAVE COVERAGE

$114.75

$27.95

FOR EASY TIME PAYMENT PLAN SEE PAGES 89-92

1938 Catalog Page

---

8 TUBE 3 Band *all-Wave* SUPERHET

Lafayette

MODEL D53

$36.95

TABLE MODEL

HAS TUNING "EYE"

MODERN CABINET

MODEL D-50

$28.75

1938 Catalog Page

ELECTRIC with TOUCH TUNING
Lafayette
AGAIN... IS OUT IN FRONT!

CONSOLE MODEL B34

$77.50

MODEL B33
CHASSIS AND SPEAKER ONLY

$77.50

★ Tunes By Electricity
★ Improved 12" Dynamic
★ Covers ALL Short Waves
★ Audio Compensator
★ Automat. Tone Expander
★ All-Range Transformer
★ Mystic "Eye"

THERMOSTATICALLY CONTROLLED CIRCUITS

CONTINUOUS WAVE *Superhet*

1938 Catalog Page

40

1938 Catalog Page

1938 Catalog Page

1938 Catalog Page

1938 Catalog Page

41

Four Sensational New Sonora Models
CLEAR AS A BELL

1938 Catalog Page

Model A-11

MODEL A-11—5 TUBE AC-DC T.R.F.

As Low As

8¹³ Net

- Full 2-Watt Undistorted Output
- Full Vision Illuminated Slide-Rule Dial
- Smartly Styled Bakelite Cabinet

Compare it with anything the market has to offer—for beauty, for performance, and for sheer value. Only then, will you realize what a real winner we are offering you in the Sonora Model A-11. Precision engineered TRF circuit provides a high degree of radio reception. Covers entire broadcast bands, as well as 1712 K. C. police channel. Through the use of the 25 L6G Beam power tube, an undistorted output of fully 2 watts is achieved. Full size electro-dynamic speaker of latest design. Full Vision Slide-Rule Dial with illuminated translucent scale and horizontal traveling pointer. Equipped with built-in aerial. Molded cabinet of striking, modern design, in choice of three colors. Size, 11¼" long, 6⅝" high, 5⅝" deep. Ship. wt., 7½ lbs. Complete with tubes. Operates on 110-120 volt AC or DC, 40 to 60 cycles.

No. 9H4253. Black ⎫ List $14.95. Dlr's., ea. $8.72.
No. 9H4254. Walnut ⎰ 3 lots, ea. $8.30, less 2%, net **8.13**
No. 9H4255. Ivory. List $16.95. Dlr's., ea., $10.47. Lots of 3, ea. $9.72, less 2%, net.......... **9.53**

MODEL C-22—6 TUBE AC SUPERHET

- New Dyna-Boost Circuit—2 Watt Output
- Full Vision Slide-Rule Illuminated Dial
- Distinctive Bakelite Cabinet Design

As Low As

11¹⁴ Net

Yes sir, a 6 tube Superhet, with all the sharp, natural performance that goes with it —yet, look at that price. Truly, it's amazing! Bakelite cabinet, designed in the very latest modern manner, is available in choice of three colors. Will give sharp, clear reception on all standard broadcasts, as well as the popular 1712 K. C. police channel. A newly developed engineering feature, provides 2 watts of undistorted output. Full size Electro Dynamic Speaker, adequately baffled for full volume handling capacity. Slide Rule dial with illuminated translucent scale in color and horizontal traveling pointer. Operates on 110-120 volts, 50-60 cycles A.C. Size, 11¾" long, 7¾" high, 7½" deep. Ship. wt., 11 lbs. Complete with tubes.

No. 9H4260. Black ⎫ List $22.95. Dlr's. $12.32
No. 9H4261. Walnut ⎰ 3 lots, ea. $11.37, less 2% **11.14**
No. 9H4263. Ivory. List $24.95. Dlr's., ea. $13.77. Lots of 3, ea. $13.08, less 2%, net...... **12.82**

Model C-22

Model EA-33

No. 9H4266. Model EA-33. List $39.95. Dlr's., ea. $20.97. Lots of 6, ea. $19.92, less 2%, net................. **19.52**

Model E-33—Without Automatic Tuning
Same as the model EA-33 without the automatic tuning feature.
No. 9H4265. List $34.95. Dlr's., ea. $19.48. Lots of 3, ea. $17.97, less 2%, net......................... **17.61**

Model EA-33—6 TUBE AC SUPERHET
With AUTOMATIC PUSH BUTTON Tuning

- 2 Bands—Covers Foreign Channels
- 6 Station Push Button Selector
- Full Automatic Volume Control
- Dyna-Boost Circuit—Develops 3 Watt Output
- Continuously Variable Tone Control
- Illuminated Slide-Rule Dial

Another Super-Sonora value, engineered for quality reception, styled for instant appeal and priced for volume sales. Automatic push button tuning feature is absolutely foolproof, instantaneous and precise. Permits selection of 6 favorite stations. Tunes 1720 to 535 KC and from 5620 to 18100 KC covering domestic and foreign short wave channels, as well as the entire standard broadcast band. 6" slide-rule illuminated dial with horizontal traveling pointer. Both wave bands clearly calibrated. 6" electro dynamic speaker. Full A. V. C. and continuously variable tone control. Exclusive Sonora feature, the Dyna-Boost circuit, provides 3 watt output—equivalent to 8 tube performance. Built-in wave trap blocks out code interference on broadcast band. Attractive new cabinet design, features corner type grille made of selected choicely grained walnut. Size: 17½" long, 18" deep, 10¼" high. Ship. wt. 20 lbs. Operates on 110-120 volt, 50-60 cycles AC. Complete with tubes.

MODEL P-99 "TEENY-WEENY"—4 TUBE AC-DC T.R.F.
Only 6¾ Inches Wide and Weighs Less Than 5 Lbs.

Sure it's tiny! In fact, so tiny that it can be conveniently held in the palm of a man's hand. But talk about performance and power! Engineered with all the infinite skill that characterizes the entire Sonora line, this little fellow enjoys such features as 2 Watt Beam Power Output, Electro-Dynamic Speaker, Full Vision Illuminated Dial, and Built-in Aerial. Can be carried and played most anywhere. Attractive walnut cabinet features smooth, round-molded lines and is available in choice of three finishes. Size: 6¾" long, 4⅞" high, 4⅞" deep. Wt., 4¾ lbs. Complete with tubes. Operates on 110-120 volt, AC or DC, 40 to 60 cycles.

- Precision Engineered T.R.F. Circuit
- Beam Power Output of 2 Full Watts
- Illuminated Full Vision Dial
- Built-in Aerial
- Operates on AC or DC Current

As Low As

8⁴⁶ Net

No. 9H4250. Walnut Finish. List $14.95. Dlr's., $9.07. Lots of 3, ea. $8.63, less 2%, net..... **8.46**
No. 9H4251. Ivory Finish.
No. 9H4252. Ebony Finish. List $14.95. Dlr's., ea. $9.77. Lots of 3, ea. $9.07, less 2%, net.......... **8.89**

AMAZING ECONOMY..THESE NEW 1½-VOLT SILVERTONE RADIOS CUT BATTERY EXPENSES RIGHT IN HALF

RADIO

Reduced to $24⁵⁰ CASH
Less Batteries
$3 DOWN

Reduced to $19⁹⁵ CASH
Less Batteries
$2.50 DOWN

6 TUBES ... 6 PUSH BUTTONS ... 2 BANDS

America's No. 1 Battery Radio—a de luxe performer, more powerful, more beautiful than any radio we've ever seen under $50, yet the price has been cut for this Sale! Six new economical 1½-volt type tubes that give 8-tube performance.

900-Hour Battery Life Guarantee: Nearly one year of service at 3 hours a day without battery replacement.

Foreign and American Reception: Two complete and separate tuning bands. Foreign and short wave range 6 to 18 M.C. American broadcast range, 545 to 1720 K.C.

6-Station Push-Button Tuning and other modern features: A touch of your finger instantly tunes any of your six favorite stations. Automatic Volume Control, Tone Control and a 6-inch alloy dynamic speaker.

Beautifully Proportioned Cabinet: Genuine two-tone walnut veneers with roll front. Size, 21½ x 13¼ x 12⅛ inches.

For Most Economy Order Radio With Power Pack Battery

★**57 J 6244**—Radio with enclosed 1½-volt Dry "A" and "B" Power Pack Battery. Mailable. Shpg. wt., 62 lbs..**$30.29**
★**57 JM 6245**—Radio with Storage "A" and two enclosed Dry "B" batteries. Not mailable. Shpg. wt., 68 lbs...**$30.27**
★**57 J 6263**—Radio Only (less batteries). Mailable. Shipping weight, 26 pounds...........................**$24.50**
★**57 J 5575**—Standard Aerial Kit. Shpg. wt., 2 lbs. .**59c**

5 TUBES ... 5 PUSH BUTTONS ... 2 BANDS

What a Sale! It's a $40 value! Our most popular battery radio reduced in price for this Sale! It has more power, yet costs less to operate than any other five-tube radio that we know of. Two dual-purpose tubes give powerful 7-tube performance.

800-Hour Battery Life Guarantee: The 1½-volt tubes use so little current that we are able to guarantee nearly nine months of service without battery replacement if you use your radio 3 hours a day.

Push-Button Tuning: The finest and simplest system of automatic tuning ever developed. No extra battery drain. Just push a button and get any one of your 5 favorite stations.

American and Foreign Reception: Foreign and short wave stations are easily tuned on the separate band with a range of 6 to 18 M.C. Broadcast band from 545 to 1720 K.C.

Modern Cabinet: A handsome table cabinet with a front panel of lovely butterfly walnut veneer. Size, 19⅝ x 11⅞ x 11 inches high.

For Top Economy Order Radio With Power Pack Battery

★**57 J 6242**—Radio with enclosed "A" and "B" Dry Battery Power Pack. Mailable. Shipping weight, 46 pounds..............**$23.90**
★**57 JM 6243**—Radio with Storage "A" and two "B" Batteries. Not Mailable. Shipping weight, 57 pounds................**$25.42**
★**57 J 6262**—Radio, less batteries. Shpg. wt., 22 lbs.... **19.95**
★**57 J 5575**—Standard Aerial Kit. Shpg. wt., 2 lbs. **59c**

FOR EASY TERMS—SEE PAGE 4A

5 TUBES ... 4 PUSH BUTTONS

$16⁹⁵ CASH
Less Batteries
$2 DOWN

Values like this are making millions change to Silvertone Battery Radios! Reduced for this Sale to make it the lowest priced 5-tube battery radio on the market. Think of being able to operate this powerful set 3 hours a day for nearly 9 months without replacing batteries! 800 hours of guaranteed service! 7-tube performance and a range of 545 to 1720 K.C.

A surprising and unlooked-for feature in a radio priced so low is the automatic 4-station push-button tuning. Other fine features include a 5-in. dynamic speaker, automatic volume control, and on-off indicator.

A handsome cabinet with a front panel and sides of gleaming genuine walnut veneer. Size, 17½ x 11 x 11⅜ in. deep.

★**57 J 6240**—Radio with "A" and "B" Dry Battery Power Pack. Mailable. Shipping weight, 42 lbs..**$20.90**
★**57 JM 6241**—Radio with Storage "A" and 2 dry "B" batteries. Not Mailable. Shpg. wt., 54 lbs..**$22.42**
★**57 J 6261**—Radio only (less batteries). Mailable. Shipping weight, 18 lbs. Sale Price.........**$16.95**
★**57 J 5575**—Standard Aerial Kit. Shpg. wt., 2 lbs.**59c**

PAGE 146 ✪ SEARS | ★**SHIPPING POINTS:** Radios and accessories shipped from **Atlanta, Memphis, Dallas,** or **Houston.** Order from one nearest you.

AMERICA'S No. 1 RADIO BARGAIN

$5.95 CHOICE OF 4 COLORS

- Was $6.49 In Our Big Catalog
- Black, Ivory, Walnut, or Red
- 4 Tubes—A.C. or D.C. Operation

America has fallen in love with this little radio—bought it by the thousands. You saw it in full colors on the inside back cover of our Spring Catalog.

Most amazing of all is the stirring tone quality of this tiny radio. A life-like tone that you'd never expect from so small a set—just 6¼ inches wide, 4¼ inches deep, and 4¾ inches high!

Expensive features include Easy Tuning Dial that lights up . . . Full 3½-inch dynamic (not cheaper magnetic type) speaker . . . Tuning range 550 to 1720 K.C. . . . Attached aerial hank.

For 110 to 120-volt, 50–60-cycle Alternating or Direct Current. (Not available for 25-cycle). Shpg. wt., 5 lbs. 8 oz. Mailable.

★57J6106–Black Plastic.....**$5.95**
★57J6107–Ivory Plastic..... **5.95**
★57J6108–Walnut Plastic **5.95**
★57J6116–Red Plastic...... **5.95**
★57J5557–**Black Fabric Zipper Carrying Case** for above. Shpg. wt., 9 oz. **79c**

Savings of a Lifet

1939 Catalog Page

ON 5-TUBE ELECTRIC RADI

- 5-Tube Superheterodyne
- Built-on Loop Aerial
- Underwriters Approved

$7.95 BLACK OR WALNUT

The finest tiny radio built under $12-95! And during this Sale it comes to you with a loop aerial built on the back of the case **at no extra cost.** Makes it completely portable and reduces man-made static and interference. Carry it from room to room. No ground or aerial wires of any kind to connect. Just plug in.

Unexcelled performance and outstanding volume and selectivity due to two dual-purpose tubes and one beam power output tube. Don't let the small size deceive you—it will pull in stations from all over the country between 540 and 1720 K.C.—some Police Calls.

Automatic volume control to prevent fading. Glowing illuminated dial. Full 3½-inch dynamic (not cheaper magnetic type) speaker.

For 110-volt, 50-60 cycle, A.C. or D.C. (Not available for 25-cycle). Size, 8¾x4⅞x4½ in. deep. Shpg. wt., 6 lbs. 2 oz. Mailable.

★57 J 6177A—Black Plastic........**$7.95**
★57 J 6178A—Ivory Color Plastic... **8.45**
★57 J 6179A—Walnut Plastic....... **7.95**

- 5-Tube Streamliner.
- Push-Button Tuning.
- Full 5-in. Dynamic Speaker.

$8.95 BLACK OR WALNUT

Until August 31—America's lowest priced **Automatic Tuning** radio! Our famous "Streamliner" with the beautiful cabinet that has sold 300,00(radios. And **now** with push button tuning at a very attractive price. Your choice of your fou favorite stations at the touch of a button. Tabs fo station call letters furnished.

5-inch electro-dynamic speaker produce tone that is almost unbelievable in its mellow life-like quality. Attached aerial hank. Tuning rang from 545 to 1720 K.C. gets American station and some Police Calls.

Molded plastic cabinet appears the same fron and back. Size, 7¼x5¾x9¾ in.

For 110 to 125-volt, 50 or 60 cycle A.C. or D.C (25-cycle not available.) Shpg. wt., 11 lbs. 4 oz Mailable.

★57 J 6102A—**Black Plastic**.......**$8.95**
★57 J 6103A—**Ivory Plastic**........ **9.95**
★57 J 6105A—**Walnut Color Plastic** **8.95**
★57 J 5569—**Suede cloth zipper carrying case** for above radios. Shpg. wt., 1 lb.....**$1.29**

YOU CAN BUY ON EASY TERMS . . . SEE PAGE 4A

8-TUBE ELECTRIC REDUCED!

Reduced to **$28.45** Cash

$3.00 DOWN

The finest All-Electric Table model that we sell! More quality features, more power, and finer tone than many radios costing up to $50.

6-Station Push-Button Tuning of most advanced type. Instantly gets you choice of six stations tuned perfectly. Electric Tuning Eye.

American and Foreign Reception: Two tuning bands. Range of Foreign band is 6 to 18 M.C.—American broadcast band 545 to 1720 K.C.

9-Tube Performance from 8 tubes including one dual-purpose tube. An advanced superheterodyne circuit that outperforms many radios at twice this price. Automatic volume control. Tone control.

Extra large dynamic speaker brings tone qualities to your ear that you'd expect only in an expensive console.

Distinctive Cabinet of rich sliced, stripe and stump walnut veneers hand rubbed to a gleaming luster. Size 21¾x9¼x12⅜ inches high.

For 110 to 125 volt, 50–60 cycle A.C. If wanted for 25-cycle, state and send $2.50 extra. Approved by Underwriters. Mailable.

★57 J 6252—Shpg. wt., 27 lbs. Cash Price.............**$28.45**
Easy Payment Price ($3 down, $4 a month)................ **31.05**
★57 J 5575—**Standard Aerial Kit.** Shpg. wt., 2 lbs..........**59c**

PRICE CUT . . . 7-TUBE ELECTRIC!

Was ~~$22.95~~ **$20.95** Cash

$2.50 DOWN

Tune in on bigger savings during this Midsummer Sale! Here's a big 7-tube radio worth $45 selling for **less than half** of that! Not a thing has been changed. It's the same fine radio that's in our Spring Catalog and this new price says "buy now!"

5-Station Push-Button Tuning. You get five different stations by just touching a button. Tuning eye aids ordinary tuning.

American and Foreign Reception: Gets everything on the air worth hearing—foreign stations and short wave programs on the 6 to 18 M.C. band and all American stations on 545 to 1720 K.C.

Improved Dynamic Speaker of the latest design insures perfect and realistic tone. Variable Tone Control and Automatic Volume Control.

Two-toned Cabinet with stylish rounded end. Instrument panel and end of sliced walnut veneer. Size, 18⅛x8x11¼ in. high.

For 110 to 125 volt, 50–60 cycle A.C. If wanted for 25-cycle, so state and send $2.00 extra. Approved by Underwriters Laboratories.

★57 J 6251—Shipping weight, 21 pounds. Cash Price.....**$20.95**
Easy Payment Price ($2.50 down, $3 a month)............... **22.75**
★57 J 5575—Standard Aerial Kit. Shpg. wt., 2 pounds........**59c**

★**SHIPPING POINTS:** All Radios and Accessories on these pages shipped from Atlanta, Memphis, Dallas, or Houston. Order from one nearest you.

They're New! Thrilling!

Enjoy these Latest Triumphs of Philco Research in

Glorious Tone and Superb Performance!

YOURS FOR ONLY A FEW PENNIES A DAY!

YEAR by year, the scientific achievements from the great Philco laboratories have made radio tone more gloriously real, performance more thrilling and the joy of fine radio entertainment more easily within the reach of every home. Today, the owners of almost twelve million Philco radios enjoy the fruits of the inventive genius of Philco engineers. Wherever radio is known, Philco is famous for that *finer* tone and performance which is born of the millions of dollars spent in scientific research.

Whatever the purpose, whatever the price, your nearest dealer offers a Philco to suit your need. Visit him today. He'll gladly demonstrate how Philco gives you the utmost in radio pleasure for the price you wish to pay.

TRANSITONE . . . *(Left)*
Sold and Guaranteed by PHILCO

New Philco inventions, new tubes, new speaker refinements give you amazingly rich tone and powerful performance never before heard in radios of this compact size. And, for the first time, Underwriters' Laboratories, Inc. approval! No hot wires, SAFE from fire and shock, SAFE for your home and children! Several models, beginning at only $9.95, full complete price including built-in aerial.

THE TRAVELER . . . *(Right)*
"She shall have music wherever she goes!"

An entirely new kind of radio, invented by Philco engineers . . . the Philco 71T. Portable . . . self-powered . . . needs no aerial, ground or "house-current". Take it with you wherever you go . . . traveling, on trains, in hotels. Use it in camps, cottages, boats, at bathing beaches—anywhere indoors or outdoors. It plays, without "hooking up" to anything!

1939 Catalog Page

PHILCO MYSTERY CONTROL

The great Philco 116RX, above, gives you radio's finest in depth and richness of tone, in superb American and Foreign reception, in beauty of design and costly cabinet woods. In addition, it brings you Mystery Control—the most miraculous radio invention since radio itself. Remote Control, without wires or plug-in connections to radio, electric outlet or anything else. Wherever you wish in your home, you change stations, adjust volume and turn the radio off without the annoyance of jumping up and running to and fro. In many cases, your present radio may be traded in as part payment and the easiest of monthly terms arranged for the balance.

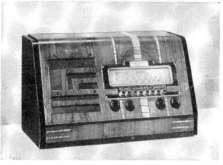

LOVELY, STREAMLINE TABLE MODELS

If you prefer fine radio tone and performance in smaller space, Philco table models are your answer! Powerful American and Foreign reception, Electric Push-Button Tuning, all the costly features of a big radio. Housed in table cabinets of handsome design in richly finished, inlaid Walnut woods. Just $5.50 down puts the 30T (illustrated) in your home!

THE NEW SPINET FURNITURE DESIGN

The new vogue in radio furniture, acclaimed by home decoration experts everywhere. Exquisite simplicity . . . graceful styling that adds to the beauty of your home, that does not clash with its surroundings. Philco Spinet style models give you the finest performance-features including Electric Push-Button Tuning for 8 stations. $10 down puts the 40XX (illustrated) in your home!

PHILCO
A Musical Instrument of Quality

PHILCO ALSO OFFERS YOU—

a complete selection of Battery Radios for Farm Homes, Phonographs with Philco, Philco Auto Radios, Philco High-Efficiency Radio Tubes and the new *vertical* Philco Safety Aerial.

6-TUBE, 2-BAND, 1½-VOLT BATTERY CONSOLE MODEL

With Economical 1½-Volt Tubes and Powerful 6-Tube, Super-Heterodyne Circuit. Assures low upkeep cost, marvelous beauty of tone and fine performance. Actually 600 hours dry battery service Guaranteed when used 3 hours daily (equal to 6 full months).

Powerful 8-inch Alloy Dynamic Speaker with 12-inch Projectotone. Volume and Tone Control. Automatic Tuning for six stations.

Gets Both Foreign and American Stations on extended range (540 to 1730 K.C. and 5.7 to 18.0 M.C.). New Drum Dial with push-button light makes tuning much easier—more accurate.

Attractive Walnut Veneer Cabinet hand rubbed to a smooth silky gloss. Beautiful Walnut Veneer top and graceful sides accented with smart butterfly figuring. Size: 24½ by 12½ by 40 inches high. *Complete* with 6 Super Airline Tubes, Dry or Storage "A" Battery, two heavy-duty "B" Batteries and Instructions. Aerial extra, see Page 665. *Shipped from* Baltimore, Albany or Pittsburgh. Send your order to nearest House. *Not Mailable.*

$**42**^{95}$ Cash
$5 Down, $5 a Month

P162 C 2660—With Dry "A" Battery. Ship. wt. 82 lbs. Cash Price...........$42.95
Time Payment Price: $5 Down, $5 a Month..................................$46.75
P162 C 1660—With Storage "A" Battery. Ship. wt. 90 lbs. Cash Price........$43.95
Time Payment Price: $5 Down, $5 a Month..................................$47.85

5-TUBE, 2-BAND, 1½-VOLT BATTERY CONSOLE GRANDE

Compare the performance, beauty and economy of this 1½-volt Battery Radio with any other 5-tube battery set. That's the best way to prove it's best! Dry "A" Battery Guaranteed for 650 Hours when used 3 hours daily.

You'll thrill to it's unequalled cabinet beauty—its lovely lines—its shiny mirror finish. Made of the finest selected Walnut Veneers throughout, with an attractive matching heart figure on dial panel. All batteries fit in hand rubbed cabinet (27¼ by 10½ by 35 in. high). Enjoy hearing American and Foreign stations (Range: 540 to 1730 K.C. and 5.7 to 18.0 M.C.). New Drum Dial makes tuning easier, more accurate.

Instant 6 Station Automatic Tuning with Automatic Volume Control brings you quick, instant tuning for your favorite stations. Tone Control gives you your exact tone preference from mellow bass to highest treble. 8-inch Alloy-Dynamic Speaker with 12-inch Projectotone. *Complete* with Tubes, one Dry or Storage "A" Battery, two "B" Batteries, Instructions. Aerial extra, see Page 665. *Shipped from* Baltimore, Albany or Pittsburgh. Send order to nearest House. *Not Mailable.*

$**36**^{95}$ Cash
$4 Down, $5 a Month

P162 C 2561—With Dry "A" Battery. Ship. wt. 75 lbs. Cash Price....................$36.95
Time Payment Price: $4 Down, $5 a Month.....................................$40.25
P162 C 1561—With Storage "A" Battery. Ship. wt. 83 lbs. Cash Price................$37.95
Time Payment Price: $4 Down, $5 a Month.....................................$41.35

6-TUBE, 2-BAND, 1½-VOLT BATTERY MANTEL MODEL

The same fine radio described above in a handsome walnut Veneer table cabinet. 6-inch Alloy Dynamic Speaker assures good tone reproduction. Plus the many other features described above. Size: 23½ by 11¼ by 11¾ in. high. Automatic Tuning for 6 stations. 600-hour dry battery guarantee when used 3 hours daily. *Complete* with 6 Super Airline Tubes, one Dry or Storage "A" Battery, two Heavy Duty "B" Batteries and Instructions. Batteries fit in cabinet. Aerial extra, see Page 665. *Shipped from* Baltimore, Albany or Pittsburgh. Send order to nearest House.

$**33**^{95}$ Cash
$4 Down, $5 a Month

P362 C 2659—With Dry "A" Battery. *Mailable.* Cash Price....................$33.95
Time Payment Price: $4 Down, $5 a Month...........................$36.95

| 656 WARDS BA | P162 C 1659—With Storage "A" Battery. *Not Mailable.* Cash $34.95 |
| | Time Payment Price: $4 Down, $5 a Month...........$38.05 |

5-TUBE, 2-BAND, 1½-VOLT BATTERY MANTEL MODEL

All the fine features of the Console Grande fully described above in a delightful table cabinet. Smooth glossy finish walnut veneer top with front and ends of walnut veneer with striking Blackwood bands. No protruding battery wires; batteries fit in cabinet. Dry "A" Battery Guaranteed for 650 hours. Size: 22½ by 11 by 11¾ inches high. Powerful six-inch Alloy Dynamic speaker for life-like tone. See Home Trial Offer on opposite page. *Complete* with 5 Super Airline Tubes, one Dry or Storage "A" Battery, two "B" Batteries and simple instructions. Aerial extra (see Page 665). *Shipped from* nearest House.

$**25**^{95}$ Cash
$3 Down, $4 a Month

P362 C 2563—With Dry "A" Battery. Ship. wt. 48 lbs. *Mailable.* Cash Price..........$25.95
Time Payment Price: $3 Down, $4 a Month.................................$28.25
P162 C 1563—With Storage "A" Battery. Ship. wt. 56 lbs. *Not Mailable.* Cash Price...$26.95
Time Payment Price: $3 Down, $4 a Month.................................$29.35

WE'RE NOT THE ONLY ONES WHO SELL GOOD RADIOS BUT

EVERY WARD RADIO SPEAKS FOR ITSELF

Prove the Quality, Beauty, Performance and Saving with a 15-Day Trial in Your Home

Hear for Yourself the Tone That Is Proof of Quality

Order any Ward Radio in the regular way (Cash or Down Payment). Try it for 15 days. Compare Quality, Beauty, Performance with radios selling for two or even three times its price. If you're not convinced it's the best dollar-for-dollar radio value on the market, send it back; we'll refund every cent and all shipping charges.

QUALITY: "Talking daily over the two-way radio to T.W.A. pilots flying passengers, mail and express between New York and California is my job. Naturally when I chose a radio, my experience told me the new Ward 'Airline' gave me the most for my money," says J. M. Sigvaldson, Chief Radio Operator, Transcontinental & Western Air, Inc.

BEAUTY: "I've seen thousands thrilled by the beauties of the ever-changing sky line and landscape during my many hours in the air, so I've come to know and appreciate real beauty. That's why I chose a Ward 'Airline' Radio. They're so attractive in design and finish," says Miss Olga Harbaugh, Chief Hostess, Transcontinental & Western Air, Inc.

PERFORMANCE: "In more than 1,500,000 miles in the air, I've carried passengers, mail and express on all sectors of T.W.A.'s 'Sunny Santa Fe Trail' airway and I have learned to judge the merits of a good radio. That's why I chose a Ward 'Airline' for my home," says Capt. Otis F. Bryan, Chief Pilot, Transcontinental & Western Air, Inc.

SAVING: The fact that Wards sell direct from factory-to-Wards-to-You is the reason our prices are so far below other Nationally Advertised sets. While others spend millions in advertising and distribution costs, Wards sell direct by mail and pass these many savings on to you. That is why you can save one-third to one-half by buying from Wards.

BEAUTIFUL 6-TUBE A.C. — D.C. RADIO WITH $13⁹⁵ Cash

6-Station Automatic Tuning — Electric Eye — Brown or Ivory Finish

Looks and Sounds Like a $20 Radio. Power to bring you a choice of the Nation's finest radio-programs—police calls from many cities. *New, Lighted Eye-Level Drum Dial* is easier to see. *Automatic Tuning* on 6 Stations. Simply press a button for your favorite stations—tuned in perfectly. *New Flush Controls and Electric Eye* make accurate tuning even easier. Shadow in eye narrows when stations are properly tuned-in—widens when stations are not in tune. *6-tube Super Heterodyne Circuit* with built-in Loop Aerial. Also has connection for outside aerial. Automatic Volume Control. 5-inch Super-Dynamic Speaker. *Complete* with Tubes and Instructions. Size: 11⅜ by 6½ by 8 inches high. Range, 528 to 1730 K.C. For use on 105 to 125-volt A.C. or D.C. Circuit. *Shipped from Baltimore, Albany or Pittsburgh.* Send order to nearest House. (25-Cycle A.C. model $1 extra and shipped from Chicago Factory.)

$2 Down, $2 Month

P462 C 602—**Brown Finish.** Ship. wt. 15 lbs. *Mailable.* Cash Price $13.95
Time Payment Price: $2 Down, $2 a Month $15.15
P462 C 603—**Ivory Finish.** Ship. wt. 15 lbs. *Mailable.* Cash Price $15.50
Time Payment Price: $2.50 Down, $2 a Month $16.80

6-TUBE A.C.—D.C. ELECTRIC RADIO WITH $18⁵⁰ Cash

Foreign and American Wave Bands, Automatic Tuning, Tone Control

Know the Thrill of foreign reception. Listen to news and music direct from Europe and South America, in addition to the finest entertainment in America and Police Calls from many cities. (Range 528 to 1730 K.C. and 2.2 to 6.5 M.C.)
Enjoy the Convenience of Automatic Tuning for your six favorite stations. Electric Tuning Eye, new Drum Dial and Flush Controls that make tuning even easier, more accurate.
Two-Position Tone Control so you can emphasize the mellow bass or brilliant high notes. Built-in Loop Aerial. Also has outside aerial connection for distant reception. 5-inch Super-Dynamic Speaker. *Complete* with Tubes and instructions. 12⅜ by 6½ by 7¾ ins. For 105 to 125-volt A.C. or D.C. *Shipped from Baltimore, Albany or Pittsburgh. Order from nearest House.* (25-cycle A.C. Model $1 extra; shipped from Chicago Factory.) Ship. wt. 17 lbs. *Mailable.*

$2.50 Down, $3 a Month

P462 C 604—**Beautiful Brown Plastic Finish.** Cash Price $18.50
Time Payment Price: $2.50 Down, $3.00 a Month $20.10
P462 C 605—**Modern Ivory Plastic Finish.** Cash Price $19.95
Time Payment Price: $2.50 Down, $3.00 a Month $21.70

BA WARDS 651

NEW 7-TUBE RADIO-PHONOGRAPH

At the Usual Price of One Alone

$**39**⁹⁵ Cash

$4 Down, $5 a Month

The music you want when you want it. Plan your entertainment—symphony, opera or modern music. Enjoy Foreign or American radio programs or your favorite recordings at a snap of a switch.

R.C.A. Electric Phono Mechanism with synchronous motor for constant speed. An essential feature for perfect reproduction. Shock-proof mounting with soft rubber spindle cup that prevents vibration. 7-inch turn-table for playing 8, 10, 12-inch records. *R.C.A. Crystal Pickup* gives equal tone perfection from outside to center of record.

Radio is same as beautiful Walnut model P462 C 715 described on Page 652, with all of its features. Size: 20½ by 11 by 12¾ inches high.

Large 6-inch Super Dynamic Speaker produces full tone range of both radio and recordings . . . almost like listening to the artists themselves.

Complete with Super-Airline Guaranteed Tubes and Instructions. Aerial, records and needles extra, see Pages 665, 680 and 681. For use on 105 to 125 volt, 60 cycle, A.C. only. *Shipped from* Baltimore, Albany or Pittsburgh. Send order to nearest House. Ship. wt. 36 lbs.

P462 C 717—Radio-Phonograph. *Mailable*. Cash Price $39.95
Time Payment Price: $4 Down, $5 a Month $43.55

8-TUBE DELUXE RADIO-PHONOGRAPH

With Automatic Record Changer — Foreign and American Wave Bands

$**91**⁹⁵ Cash

$9 Down, $8 a Month

Beautiful Console Grande Model that brings you the very finest of both radio and phonograph music.

Carefully Selected matched stump Walnut Veneer front panel with Bubinga (rare tropical wood) bands. Sliced Walnut top, sides and pilasters—all hand rubbed till it shines like a mirror. Self balancing hinges hold top open in any desired position. Two convenient compartments in which to store 10 or 12-inch records or record books.

DeLuxe Automatic Record Changer plays up to twelve 10-inch or ten 12-inch records automatically . . . 30 minutes of carefree entertainment. Any record may be rejected at will, and last record is repeated automatically. Self-starting electric motor, precision made for smooth, quiet operation and years of dependable service. Uniform speed to assure finest possible tone perfection.

High Fidelity Piezo Crystal Pickup for utmost richness and realism of tone. Mounted on tangent arm, needle is maintained at correct angle from beginning to end of record. Swings upward, too, for easy needle changing.

Prove Its Superiority in Your own Home. Order this or any Ward Radio in the regular way (Cash or Down Payment). If it isn't by far the best value for the money you can find, send it back and we'll refund every cent including all shipping charges.

Powerful 8-tube Super Heterodyne Radio with both Foreign and American Wave Bands to bring you the finest programs on the air. Automatic Tuning on 6 stations, big 10-inch Super-Dynamic Speaker, Automatic Volume Control, Electric Tuning Eye, Personal Tone Control and all the other fine features of 8-tube Radio P162 C 801, described on Page 652. Size: 32⅜ by 17¼ by 34¾ inches high.

Complete with Tubes and Instructions. Aerial, records and needles extra, see Pages 665, 680 and 681. For use on 105 to 125 volt, 60 cycle, A.C. only. *Shipped promptly from* stock. Ship. wt. 110 lbs. *Not Mailable.*

162 C 805—Radio-Phonograph. Cash Price $91.95
Time Payment Price: $9 Down, $8 a Month $100.25

TRY TO MATCH THIS GREAT RADIO BUY!

$**5**⁷⁵ Brown

Compact New 4-Tube A.C.-D.C. Super Heterodyne, Fully Licensed

A midget Radio that's more than a mere novelty. Every necessary detail perfected for *real* radio performance. Meets Airline's quality standards because it's fully licensed by R.C.A. and Hazeltine. An unusual value and real bargain at this low price.

Super Heterodyne Circuit with 4 full working tubes. Gives ample volume on local or powerful nearby stations. Tuning Range (540 to 1650 K.C.) covers all American broadcasting stations. *Automatic Volume Control* holds volume constantly where you want it!

Full-Vision Dial—Easy to read, with handy sliding pointer for quick, accurate tuning.

4-inch Permanic Speaker provides unusually pleasing tone. *Volume Selector* enables you to turn volume up to fill your room or down to a mere whisper. Approved by Radio Manufacturers' Association and Underwriters' Laboratories. *Complete* with Tubes and Instructions. Size: 7 by 4 by 4⅞ in. high. For use on any 105 to 125 volt, D.C. or 50 to 60-cycle A.C. Current.' *Shipped from* Baltimore, Albany or Pittsburgh. Send order to nearest House. For Convenient Time Payments see Page 1009. Ship. wt. 9 lbs. *Mailable.*

650 WARDS BA
P462 C 420—Handsome Brown Plastic with Louvre Style Grille. $5.75
P462 C 421—Ivory Plastic with Louvre Style Grille $6.55

CONVENIENT 5-TUBE A.C.-D.C. ELECTRIC RADIO

$**8**⁹⁵ Brown

With Push Button Automatic Tuning on Your Four Favorite Stations

More Beauty, Performance and Convenience than you can usually buy anywhere else at this price . . . actually compares feature for feature with other sets selling up to $15.

Get Your Favorite Programs Instantly simply by pressing a button. Faster, easier and more accurate than ordinary tuning. Call letters easily inserted from complete list furnished. Approved by R.M.A. and Underwriters' Laboratories; fully Licensed by R.C.A. and Hæzeltine for your protection.

Lighted Drum Shape Dial with a sliding pointer. Easier to read than the usual flat dial. Covers all American Broadcasting Stations (540 to 1650 K.C.). Full 5-inch Super-Dynamic Speaker assures rich, life-like tone. Automatic Volume Control prevents bothersome fading and blasting.

Rich Brown or Attractive Ivory Finish with back enclosed. Ideal for Student away at school. With built-in Loop Aerial for nearby stations. Also has outside aerial connection for greater distance.

Complete with Tubes, Instructions. Size: 10⅜ by 5½ by 6¾ in. high. For use on 105 to 125 volt, A.C. or D.C. Circuit. *Shipped from* Baltimore, Albany or Pittsburgh. Send order to nearest House. (25-cycle A.C. Model 50c extra and shipped from Chicago Factory stock.) Ship. wt. 10 lbs. *Mailable.*

P462 C 508—Mottled Brown Plastic . . . . $8.95   P462 C 509—Smart Ivory Plastic $10.45

WHY WAIT FOR THE HIGHLINE

TO ENJOY CITY RADIO POWER?

Why wait for the Highline when you can enjoy one of these new One-Battery Radios now—they're almost as convenient and economical as an all-electric set. One 6-volt Storage Battery supplies all the power and, in addition to your radio, it will operate electric lights, shaver, etc. Then, when the Highline does come, most of these radios are easily converted to A.C. electric models.

One 6-Volt Storage Battery supplies all the power needed. No "B" or "C" Batteries to replace! Lasts for weeks on each charge. Can be recharged by any service station or kept charged with a Wind or Gasoline Engine Charger.

Automatic Tuning on your favorite stations. Easily set for the six stations of your choice. All American Call Letters included. Does not interfere with regular tuning in any way.

Electric Eye for Perfect Tuning. When you are tuned for best reception, shadow in Eye is at its narrowest. If it widens, you are tuning away from the station you want. Very convenient and helpful.

Powerful Dynamic Speakers for marvelous richness and realism of tone... so natural and lifelike it's almost like listening to the artists themselves.

New Lighted Drum Dial, exclusive with Wards! More accurate because numbers are large enough for easy reading; spaced for more exact tuning. Flush Tuning and Volume Controls.

Order any Ward Radio in the regular way, Cash or Down Payment. Try it for 15 days. If you're not completely satisfied, send it back and we'll refund every cent including shipping charges both ways.

5-TUBE, ONE-BATTERY RADIO WITH BUILT-IN A. C. POWER UNIT

Just plug it in to the A.C. electric lighting socket or wall plug, and it's all ready for 110-volt A.C. Electric operation . . . Connect it to your 6-volt Storage Battery, and it's ready for battery operation. Nothing more to buy . . . No switches to turn . . . No complicated wiring changes to make . . . the right way is the only way it can be connected.

Automatic Tuning. Quicker, easier and more accurate than tuning by hand. Simply press a lever for your favorite programs. Does not interfere with ordinary tuning. Easily set for any 6 stations.

Illuminated, Horizontal Full-Vision Dial with easy-to-read numbers. Especially convenient because you can see the entire number scale and locate stations easily. Extended Tuning Range (540 to 1720 K.C.). Covers all American Broadcasting stations and Police Calls from many cities.

Automatic Volume Control to prevent bothersome fading. Just turn the Volume up or down till it's as loud or soft as you want, and it will stay that way!

6-inch Alloy Dynamic Speaker for clear, natural tone and ample volume. *Unsurpassed Beauty* of design and finish for a radio selling at this low price. Made of carefully selected wood with imitation Walnut finish. Harmonizing knobs, escutcheon and grille cloth. Size: 14¾ by 6¾ by 8¾ inches high. Licensed by R.C.A. and Hazeltine, and Approved by Radio Manufacturer's Association.

Complete with Super Airline Tubes and Instructions. Storage Battery and Aerial extra, see Pages 660 and 665. Operates on one 6-volt Storage Battery or 105 to 125 volt, 50 to 60 cycle, A.C. Battery does not fit in Cabinet. See extra saving when bought with Wind or Gasoline Charger, Pages 660 and 661. *Shipped from Baltimore, Albany or Pittsburgh. Send your order to nearest House. Ship. wt. 33 lbs. Mailable.*

P462 C 564—5-tube Convertible Radio. Cash... **$21.95**
Time Payment Price: $3 Down, $3 a Month... **$23.85**

$21.95 Cash
$3 Down, $3 a Month

ECONOMICAL NEW 5-TUBE, ONE-BATTERY RADIO

Try to match the quality—the low price of this 5-Tube, one-Battery Set. Take advantage of Wards 15-Day Home Trial (See above). Actually prove in your own home that it's one of the year's best buys.

A Million Dollar's Worth of Entertainment for only a few cents a month! One 6-volt Storage Battery supplies all the power... no "B" or "C" Batteries needed. (Not convertible for A.C. electric operation).

5-inch Alloy Dynamic Speaker for clear, life-like tone and plenty of volume. Full range Volume Selector, and Automatic Volume Control to prevent bothersome fading.

Distance Getting Super-Heterodyne Circuit. Gets finest American programs and Police Calls from many cities. Range: 540 to 1720 K.C. Automatic Tuning on six stations. Does not interfere with ordinary tuning. Easily set for the stations of your choice.

Latest and Most Popular Style Cabinet of attractive Brown Molded Plastic. Fits beautifully in any room. Size: 12⅝ by 6½ by 7¾ inches high. Licensed by R.C.A. and Hazeltine to assure the latest improvements. Approved by Radio Mfrs. Assn.

Complete with Tubes and Instructions. 6-volt Storage Battery and Aerial extra, see Pages 660 and 665. Battery does not fit in Cabinet. See extra saving with Chargers, Pages 660, 661. *Shipped from Baltimore, Albany or Pittsburgh. Send your order to nearest House. Ship. wt. 20 lbs. Mailable.*

P462 C 562—5-Tube One-Battery Radio. Cash Price..... **$20.95**
Time Payment Price: $3 Down, $3 a Month........... **$22.75**

$20.95 Cash
$3 Down, $3 a Month

BA WARDS 659

Sonora Celebrates!
With the JUBILEE Model P-101
4 TUBE AC-DC

As Low As $5.98

Designed with all the precision of a fine watch, Sonora engineers have succeeded in giving the "Jubilee" model an abundant measure of the sharp tonal beauty for which Sonora is famed. Its unique compactness makes it ideal for use as an auxiliary or traveling radio, as it can be easily packed in smallest luggage.

Fully 2 Watts of undistorted output is achieved thru the use of 25L6GT Beam Power Pentode Output tube. This gives it a performance quality equal to that of sets many times the size and price of the "Jubilee."

Full Vision Clock Type Dial, with scale illuminated from above. Tunes from 550 to 1720 K.C., covering Standard broadcasts and the popular 1712 K.C. POLICE CHANNEL.

Strikingly Beautiful Cabinet of molded bakelite, in rich mottled walnut effect. Measures only 6⅝" wide, 4⅝" high, 4¾" deep. Weight, 4 lbs.

Built-in Aerial makes it constantly available for use on any A.C. or D.C. current, 110-120 volts, 40 to 60 cycles. Complete with the following tubes: 1—6K7GT, 1—6J7GT, 1—25L6GT, 1—25Z6GT.

No. 9H4112. List $7.99. Dlr's., $6.85. Lots of 3, $6.10, less 2%, net........................... **5.98**

Small... But Oh! What Power!

Sonora-METRO Model P-106
4 TUBE AC-DC

As Low As $6.52

Here again, Sonora engineers and designers have accomplished the almost impossible. You will search long and far before you will encounter a midget radio within the price range of the "Metro" that can begin to compare with it in tonal quality, in undisturbed output, and in beauty of design.

Beam Power Tube gives this little model a high degree of performance quality and undistorted volume, truly remarkable for such a mite of a set. 4 latest type tubes as follows: 1—6K7GT, 1—6J7GT, 1—25L6GT, 1—25Z6GT. 2 watt output.

Tunes from 1720 to 550 K.C.s, covering the entire domestic and Canadian broadcast band as well as upper police channels. Tuning is greatly simplified by new Easy-Tune Dial. Built-in antenna. Electro-dynamic speaker.

Ivory Plastic Cabinet of extraordinary beauty has irresistible eye-appeal. Size, 6½" wide, 4½" high, 4⅝ deep. Wt. 4¼ lbs. Operates on AC or DC, 110-120 volt, 40-60 cycles.

No. 9H4100. List $9.99. Dlr's., $7.20. Lots of 3, $6.65, less 2%, net........................... **6.52**

Superbly Engineered for Quality Performance!

Sonora-COSMO Model TS-105
5 TUBE AC-DC SUPERHET

As Low As $7.69

Smartly styled—compact—and with an abundance of that Sonora quality, the Model TS-105 has all the essentials of a real sales sensation.

Full 2-Watt Output. It is difficult to imagine that such room-filling volume could emanate from this tiny little set. However, this is made possible by the powerful 25L6GT beam power pentode output tube, which provides 8-tube power and tone quality.

Gets Police Calls. Tunes from 550 to 1720 K.C. covering standard broadcasts and the popular 1712 K.C. police channel. Tuning arrangement with movable carrier on revolving disc.

8 Tube Performance. Is the result of an ingenious superheterodyne circuit actually using 5 tubes. Tubes are: 1—6A8GT, 1—6K7GT, 1—6Q7GT, 1—25L6GT and 1—25L6GT. Complete with built-in aerial. Operates on 110-120 volts, 40 to 60 cycles, AC or DC.

Exquisite Ivory Plastic Cabinet in choice of Ivory or Walnut. 8" wide, 4¾" wide, 4¾" high. Weight, 7½ lbs.

No. 9H4101. With Ivory Cabinet. List $12.95. Dlr's..$9.00. Lots of 3, $8.45, less 2%, net........ **8.28**

No. 9H4113. With Walnut Cabinet. List $11.95. Dlr's., ea. $8.50. Lots of 3, ea. $7.85, less 2%, net.. **7.69**

LICENSED UNDER R.C.A. PATENTS

1939 Catalog Page

Sonora Precision Engineered Through
Clear as a Bell

No.	Color	List	Dlr's., ea.	3, Ea.	Less 2%
9H4260.	Black	18.95	12.14	11.59	11.36
9H4261.	Walnut	19.95	12.46	11.90	11.66
9H4263.	Ivory	23.95	14.05	13.42	13.15

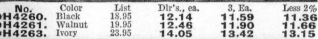

Sonora-Model C-22
6 TUBE AC SUPERHET

Featuring the DYNA-BOOST CIRCUIT

As low as

$11³⁶

One has but to see this distinctive little receiver, then listen to the smooth bell-like quality of its tone to realize at once that here is much more radio value than is indicated by the low prices at which these sets are selling.

Dyna-Boost Circuit. An exclusive Sonora feature which makes possible from two to three times more output than on ordinary sets using an equivalent number of tubes.

Full Vision Slide Rule Dial with illuminated translucent scale in color and horizontal-traveling pointer. Tunes from 535 to 1720 K. C., which covers Standard broadcasts as well as the 1712 K. C. police channel.

Superheterodyne Circuit incorporating up to the minute developments in radio design, is without an equal for razor-keen sharpness and naturalness of tone. Complete with the following R. C. A. tubes: 1—6A8G, 1—6K7, 1—6Q7G, 1—25A6G, 1—25Z6G, 1—165R4G.

Distinctive Molded Cabinets in choice of three popular colors. Entirely new honeycomb grille design, acoustically correct for the full size electro dynamic speaker. Measures 11¼" long, 6¾" high, 6½" deep. Wt. 11 lbs. Operates on 110-120 volt AC, 50-60 cycles.

8 Tube Performance Guaranteed
Sonora-Model C-221
6 TUBE AC SUPERHET

In 3 lots

$12⁴⁹
Net

- Dyna-Boost Circuit
- Full 2 Watt Output
- Illuminated Slide Rule Dial
- Tastefully Designed Cabinet

From a study of what the radio public of today is buying, we feel quite certain that this model is going to be a big volume number. Look what it offers **and at what price!** A 6 tube superheterodyne circuit with the new Dyna-Boost Feature—2 watts of undistorted output—full vision illuminated slide rule dial—full size electro dynamic speaker. Tunes from 535 to 1720 K. C. covering Standard broadcasts and the 1712 police channel.

Handsome two-tone solid walnut cabinet, in conservatively modern design. Measures 12" wide, 7⅞" high, 7½" deep. Wt. 12 lbs. Complete with the following R. C. A. tubes: 1—6A8G, 1—6K7, 1—6Q7G, 1—25A6G, 1—25Z6G, 1—165R4G. Operates on 110 volt, AC, 50-60 cycles.

No. 9H4262. List $21.95. Dlr's., ea. $13.45. Lots of 3, ea. $12.75, less 2%, net **12.49**

Dual Wave Bands..Gets Foreign Reception

Sonora-Model DD-14
6 TUBE AC-DC SUPERHET

In 3 lots

$15⁹⁷
Net

Literally brimming with quality features, this handsome Sonora table model is bound to meet with widespread popularity among radio buyers who are quick to recognize value.

Two Full Bands bring the listener a host of interesting domestic and 49-meter foreign short wave programs as well as the regular standard broadcast. Tunes from 535 to 1720 K. C. and from 5650 to 18,100 K. C.

6" Full Vision Side Rule Dial with illuminated colored scales with horizontal traveling pointer. Calibrated in kilocycles for broadcast, in megacycles for short wave. Important station positions indicated.

Superbly Engineered in every detail of its mechanical construction, the Model DD-14 delivers 8 tube performance from 6 tubes arranged in a superheterodyne circuit which features Automatic Volume Control, Full 2-watt Output, Built-in Aerial and 5-in. Electro Dynamic Speaker.

Attractively Styled Cabinet with front face and sides of one-piece roll effect in horizontally striped walnut. Overlay trims of contrastingly grained walnut finish. Distinctive grille design. Measures 15½" long, 7" deep, 8½" high. Wt. 14 lbs. Operates on 110-120 volts, 40 to 60 cycles AC or DC. Comes complete with the following R. C. A. tubes: 1—6A8G, 1—6K7, 1—6Q7G, 1—25L6G, 1—25ZL6, and 1—165R4G.

No. 9H4273. List $29.95. Dlr's., ea. $17.70. Lots of 3, ea. $16.90. Lots of 6, ea. $16.30, less 2%, net **15.97**

Licensed Under R. C. A. Patents

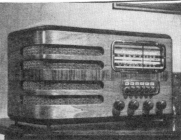

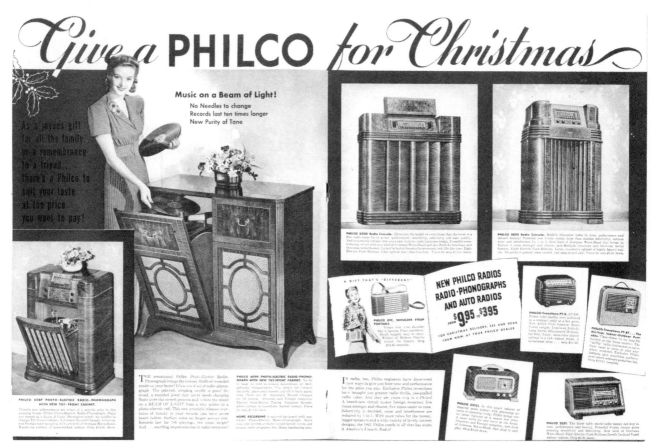

54

Sculptured PLASTICS ☆
LAFAYETTE TINY TUNERS

5 TUBE AC-DC SUPER

H ERE's your chance to get that open set you've always wanted—and at a price far below what you ever expected to pay! This sculptured performer is at home in the house, out on the porch, bedroom, den, library or any other "anywhere" place and let's provide you with good steelmanship. It has very lovely good-looking tone appearance natural and looks...isn't there's power to spare...

"Antenn-Aire" Gets All The Stations

No need to bother with connecting wires when you want that set. The magic "Antenn-Aire" eliminates the need for either aerial or outside connections, and it actually "pulls in" all the stations you want to hear. You don't see it, but it works for you silently and efficiently...

Crystal Clear "Beam" Power

Model BB-22 In Walnut Case
YOUR COST..................**$8.95**

Model BB-23 In Ivory Case
YOUR COST..................**$9.95**

6 TUBE AC-DC SUPER

Y OU'LL enjoy the convenience and luxury of push-button tuning...

Sounds Just Like A "Big Set"

Model FE-5 In Walnut Case
YOUR COST..................**$11.95**

Model FE-6 In Ivory Case
YOUR COST..................**$12.95**

WITH THE MAGIC ANTENN-AIRE — NO WIRES TO ATTACH — JUST PLUG IN

PUSH-BUTTON TUNING

1940 Catalog Page

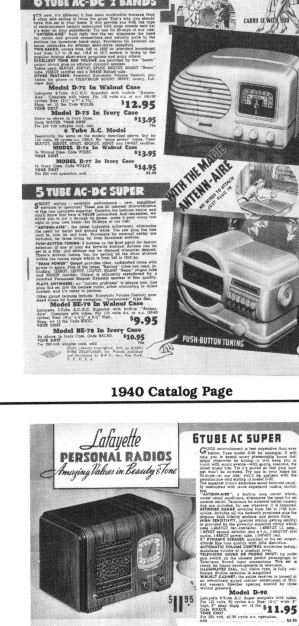

GEMS OF BEAUTY ☆
WITH AMAZING FULL-BODIED TONE

6 TUBE AC-DC 2 BANDS

CARRY IT WITH YOU

Model D-72 In Walnut Case
YOUR COST..................**$12.95**

Model D-73 In Ivory Case
YOUR COST..................**$13.95**

6 Tube A.C. Model

MODEL D-76 In Walnut Case
YOUR COST..................**$13.95**

MODEL D-77 In Ivory Case
YOUR COST..................**$14.95**

5 TUBE AC-DC SUPER

WITH THE MAGIC ANTENN-AIRE — NO WIRES TO ATTACH — JUST PLUG IN

Model BE-79 In Walnut Case
YOUR COST..................**$9.95**

Model BE-78 In Ivory Case
YOUR COST..................**$10.95**

PUSH-BUTTON TUNING

1940 Catalog Page

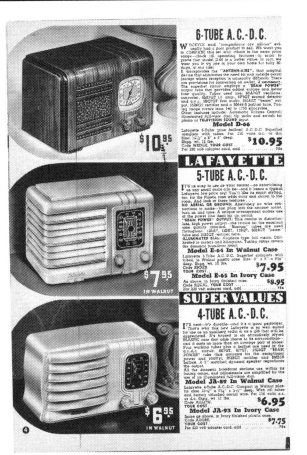

6-TUBE A.C.-D.C.

Model D-66
YOUR COST..................**$10.95**

LAFAYETTE
5-TUBE A.C.-D.C.

Model E-64 In Walnut Case
YOUR COST..................**$7.95**

Model E-65 In Ivory Case
YOUR COST..................**$8.95**

SUPER VALUES
4-TUBE A.C.-D.C.

Model JA-92 In Walnut Case
YOUR COST..................**$6.95**

Model JA-93 In Ivory Case
YOUR COST..................**$7.75**

1940 Catalog Page

Lafayette
PERSONAL RADIOS
Amazing Values in Beauty & Tone

6 TUBE AC SUPER

Model D-90
YOUR COST..................**$11.95**

5 TUBE AC-DC 2 BAND

Model JA-87 In Walnut Case
YOUR COST..................**$10.95**

Model JA-84 In Ivory Case
YOUR COST..................**$11.45**

1940 Catalog Page

55

You're set for the future with this 3 IN ONE radio!

5 TUBE 2 BAND SUPER

Model E-69

Lafayette 5-Tube Superhet for battery, 110 volt a.c. or d.c. operation. With tubes.

$24.95

Lafayette

2 Way radio versatility for the low cost of ONE!

5 TUBE 6V. DC OR 110V. AC

Model FE-35

$22.95

Lafayette

7 TUBE 6 VOLT 3 BAND SUPER

Model CC-1

$29.95

1940 Catalog Page

LAFAYETTE PERFORMANCE CASED IN *Jewel-like* **CATALIN**

THE MIRACLE PLASTIC

SPECIAL 5 TUBE AC·DC

PUSH BUTTON TUNING

Lafayette **5 TUBE '3 IN 1' RAMBLER**

WITH THE MAGIC ANTENN-AIRE

Model E-62 In Onyx Catalin Case
Model E-63 In Ivory Catalin Case **$13.95**

Model T-56 "Rambler" **$14.95**

1940 Catalog Page

DE LUXE *Lafayettes*

LABORATORY TESTED · CUSTOM BUILT · HIGH FIDELITY

Outstanding 12 TUBE AC·DC THREE BAND SUPERHET

MANTEL MODELS
Model CC-47 **$57.50**
CC-48 Long-Wave Model **$57.50**

AUTOMATIC PHONO-RADIO COMBINATION

Model CC-41 Aut. Combination **$119.95**
Model CC-46 With Single Record Player **$107.50**

SPECIAL LONG WAVE SETS
Model CC-68 Aut. Combination **$119.95**
Model CC-61 With Single Record Player **$107.50**

SEE PAGE 72 FOR FULL DETAILS ON GARRARD.

1940 Catalog Page

SOUND & PHONO EQUIPMENT

EMBLEM Record Players · **Grip-Cap Phono Adapter** · *Lafayette* **Hearing Aids "Theatrear"**

Record Player Adapter

Mike & Phono Adapter 54¢

Phono-Adapters, With Switch

Handi-Way Record Rack $7.95

QUAM Permanic Mike

RCA Record Player Connecting Switch and Cable 59¢

SPECIAL! *Lafayette* **Portable P.A. System** $19.95

LINGUAPHONE SYSTEM Learn Code the Easiest Way!

We Will Not Knowingly Be Undersold

1940 Catalog Page

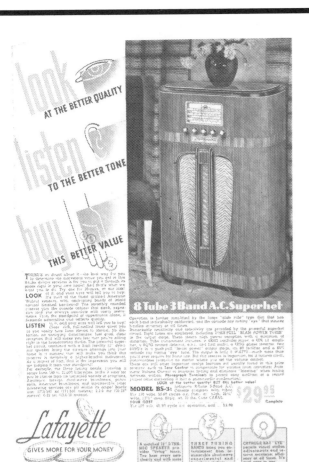

look
AT THE BETTER QUALITY

listen
TO THE BETTER TONE

buy
THIS BETTER VALUE

THERE'S no doubt about it—the best way for you to determine the increasing value you get in this 8-tube deluxe receiver is for you to put it through its paces right in your own home! And that's what we want you to do. Try one for 30 days. If it meets with your approval and your eyes will tell you to buy.

LOOK at the fine quality of the finest grained American Walnut veneers, with matching bands of inlaid genuine limewood burl-wood. The smoothly rounded corners give the console cabinet that sleek, separation look that everyone associates with costly instruments. Notice the styling throughout.

LISTEN to the better tone that will make you think this is a more expensive set. And then carefully note how it responds in the home-operating studio. The powerful superhet circuit combined with a high fidelity 12" dynamic speaker bring the listener utmost listening joy.

For example, the 3 extended wave bands traveling a range from 540 to 21,000 kilocycles make it easy for you to tune-in on unusual broadcasts.

LOOK at the better quality! 8½" WS, better value!

8 Tube 3 Band A.C. Superhet

Operation is further simplified by the large "slide rule" type dial that has each band individually calibrated, and the cathode ray tuning "eye" that insures hair-line accuracy at all times.

Remarkable sensitivity and selectivity are provided by the powerful superhet circuit. Eight tubes are employed including PUSH PULL BEAM POWER TUBES in the output stage. These insures high power reception with a minimum of distortion.

MODEL BS-3:

Console complete with tubes. For 110 volts, 50-60 cycle a.c. Ship. wt. 70 lbs. Code CARAS.

YOUR COST .... **$29.95** Complete

For 220 volt, 40-60 cycle a.c. operation, add ... $3.00

Lafayette
GIVES MORE FOR YOUR MONEY

Lafayette AUTOMATIC
PHONO RADIO COMBINATION

8 TUBE 3 BAND A.C. SUPERHET
PLAN YOUR OWN CONCERTS

BRINGS YOU EVERYTHING YOU WANT TO HEAR...THE WAY YOU WANT TO HEAR IT!

WITH GARRARD CHANGER *Automatic*

CUSTOM-BUILT TO EXCEL

FOR RADIO PROGRAMS

FOR RECORDED PROGRAMS

MANUAL PLAYER AVAILABLE

Model BB-7 Combination

YOUR COST **$99.95**

Model BB-6 Combination

YOUR COST **$84.95**

Model 996 Cabinet Only

YOUR COST **$29.95**

MODERN RADIO *belongs to you*

12" DYNAMIC SPEAKER

MAGIC ANTENN-AIRE

8 TUBE A.C. SUPER

These glass and five metal tubes are employed in a highly efficient superheterodyne circuit that is licensed under both RCA and Hazeltine patents. This circuit provides a high degree of sensitivity and selectivity and excellent fidelity. The receiver is also approved by the Underwriters.

3 BAND TUNING

Three extended wave bands are covered including:
7,000 to 21,000 KC (42.8-13.7 meters)
2,300 to 7,300 KC (136.5-42.8 meters)
539 to 1,720 KC (556.1-173.8 meters)
All of these bands are instantly available at the flip of a switch. All bands can be received on the built-in Magic Antenn-Aire aerials!

DUAL ANTENN-AIRE

Not one—but two Magic Antenn-Aires are built into this console.

$49.95

Model BB-4 Lafayette 8-Tube Console

Model BB-4

Lafayette Dual-Magic "Antenn-Aire" 3-Band, 8-Tube Superhet Console complete with tubes. Overall dimensions 20" x 11¼" x 38" high. For 110-120 volts 50/60 cycles a.c. operation. Shpg. wt. 68 lbs. Code BISON.

YOUR COST **$49.95**

For 220 V. 40/60 cycle a.c. add $2.48

YOU'LL BE PROUD TO OWN A *Lafayette*

LAFAYETTE PORTABLES
THEY PLAY *Wherever you go*

MAGIC ANTENN-AIRE · SELF-POWERED · NO WIRES TO ATTACH · JUST TUNE IN · INSTANT OPERATION

5-TUBE "3-in-1" PORTABLE

SPECIAL: ALSO OPERATES ON 110V. AC or DC

Model E-80

Lafayette 5-Tube "3-in-1" Tourist Portable in airplane luggage case with drop front. Size: 11½"x11½"x7½" deep. With tubes.

YOUR COST, LESS BATTERIES **$19.95**

K201AL—Kit of batteries for above.
YOUR COST **$2.65**

5-TUBE DELUXE PORTABLE

Model BB-70

Lafayette "Rover" 5-Tube Battery Portable in airplane luggage case with "slick' alk" design. With tubes. Shpg. wt. 21 lbs. Code SABLE.

YOUR COST, LESS BATTERIES **$16.95**

K201AA—Kit of batteries for above. $2.45

4-TUBE LIGHTWEIGHT

Model S-50

Lafayette "Petite" 4-Tube Lightweight Battery Portable complete with tubes. Shpg. wt. 17 lbs. Code PETAL.

YOUR COST, LESS BATTERIES **$11.95**

K201AA—Kit of batteries for above. $3.50

NOTE: Although our famous 30-day home trial guarantee applies to all Lafayette battery sets, batteries themselves (for obvious reasons) are not subject to return for credit or refund.

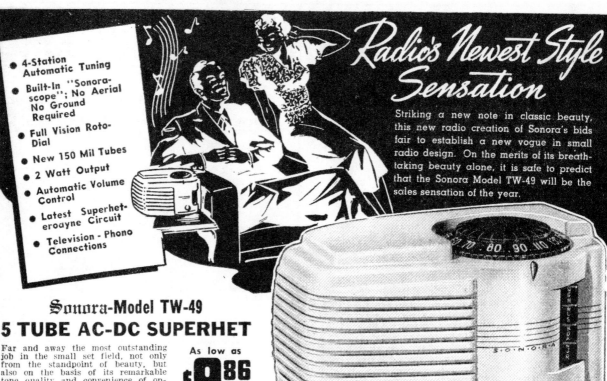

- 4-Station Automatic Tuning
- Built-In "Sonora-scope"; No Aerial No Ground Required
- Full Vision Roto-Dial
- New 150 Mil Tubes
- 2 Watt Output
- Automatic Volume Control
- Latest Superhet-eroayne Circuit
- Television - Phono Connections

Sonora-Model TW-49
5 TUBE AC-DC SUPERHET

As low as
$9 86

Far and away the most outstanding job in the small set field, not only from the standpoint of beauty, but also on the basis of its remarkable tone quality and convenience of operation. We suggest building your entire radio display around this model, because it will undoubtedly be one of the most popular numbers in the Sonora line.

4 stations can be front-adjusted for automatic push button selection with effortless ⅜" stroke. Manual tuning is made exceedingly simple by means of the 5" Roto-Dial. Tunes from 1720 to 535 K.C.'s covering the entire broadcast band and 1712 K.C. police channel.

Latest Superheterodyne circuit employs 5 new 150 mil tubes—no ballast, no heater cord. Built in "Sonora-scope" eliminates necessity for any aerial or ground connections. Tubes are as follows: 1—12A8GT, 1—12K7GT, 1—12Q7GT, 1—35L6GT, 1—35Z5. 5" P.M. Dynamic speaker.

A real achievement in design in the Model TW-49 has in its molded classic cabinet a fitting complement to the glorious tone which emanates from it. Available in choice of three attractive finishes: ivory, black and onyx. Dimensions, 11¼" wide, 6¾" high, 6½" deep. Operates on 110-120 volts, 40-60 cycles AC or DC. Ship. wt., 7½ lbs.

Automatic Tuning!

No. 9909. Walnut finish. List $15.95. Dlr's., ea. $10.58.
Lots of 3, ea. $10.06, less 2%, net............................ **9.86**

No. 9910. Ivory finish. List $17.95. Dlr's., ea. $11.93.
Lots of 3, ea. $11.47, less 2%, net............................ **11.24**

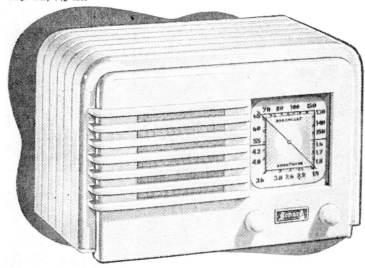

Appealing, simple, yet distinctly modern is the classic molded cabinet in which the Model TJ-62 is housed. Offered in choice of rich walnut or soft ivory finish. Unique louvre speaker grille design. Dimensions, 12" wide, 7¼" high, 7" deep. Ship. wt., 9 lbs.
No. 9912. Walnut finish. List $18.95. Dlr's., ea.
$11.79. Lots of 3, ea. $11.22, less 2%, net...................... **11.00**
No. 9913. Ivory finish. List $20.95. Dlr's.,
$13.27. Lots of 3, ea. $12.76, less 2%, net...................... **12.50**

Sonora-Model TJ-62
5 TUBE AC-DC SUPERHET

- Built-In Sonorascope— No Aerial or Ground Required
- 8-Tube Performance
- 2-Watt Output
- Television-Phono Connections
- Automatic Volume Control

As low as
$11 00

The masterful radio craftsmanship which typifies the name of Sonora is brought out to full advantage in this little 5-tube plastic model.

Superbly engineered superheterodyne circuit uses 5 new 150 mil type tubes—no ballast, no heater cord. Built-in "Sonorascope" does away with aerial or ground connections. Just plug in and tune.

Tubes from 1720 to 535 K.C.'s covering the entire broadcast band as well as the 1712 Police Channel. New Full-vision square Gem-loid dial in rich gold and cream finish. Other features include automatic volume control, 5" P.M. Dynamic speaker, television and phono connections and 2-watt output. With the following tubes: 1—12A8GT, 1—12K7GT, 1—12Q7GT, 1—35L6GT, and 1—35Z5. Operates on 110-120 volts, 40-60 cycles AC or DC.

1940 Catalog Page

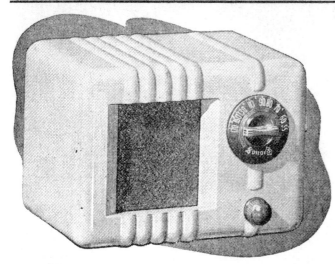

No. **9900.** Walnut finish. List $8.95. Dlr's., ea. $5.90. Lots of 3, ea. **$5.66**, less 2%, net.................... **5.55**

No. **9901.** Ivory finish. List $9.45. Dlr's., ea. $6.60. Lots of 3, ea. **$6.10**, less 2%, net.................... **5.98**

Sonora's "TEENY-WEENY" LEADER
Model TP-108 — 4 Tube AC-DC

As Low As

$5.55

The demand for tiny little receivers, such as the "Teeny-Weeny," is increasing by leaps and bounds. This attractive little model by Sonora has the tonal excellence, the smart styling and is attractively priced to garner a considerable share of the small set market for Sonora dealers. There is no doubt, but what one of these on display will enable you to dispose of a sizable quantity of this number.

A Remarkable Engineering Achievement Sonora engineers have really done a masterful job in the creation of this little 4-tube job, giving it an amazing degree of sharp selectivity and tonal beauty. Tunes from 1720 to 535 K.C. covering the entire broadcast band as well as the 1712 K.C. police channel. Full vision type dial, colored in attractive two-tone effect, highlights beauty of cabinet.

Surprisingly Powerful! Employs 4 new 150 mil tubes— no ballast, no heater cord. Tubes are as follows: 1—12K7GT, 1—12J7GT, 1—50L6GT, 1—35Z4GT. Has an exceptionally high output of 1½ watts—equal to that of high priced table model receivers. P. M. Dynamic speaker. Attached 20 ft. antenna.

Smart, Appealing Cabinet Design Presented in a striking little plastic molded cabinet of distinctive modern design, in choice of ivory or walnut finish. Open grille permits finest acoustical reproduction. Size, 6½ in. wide, 4½ in. high, 4¾ in. deep. Ship. wt., 4¼ lbs. Operates on 110-120 volts, 40-60 cycles, A.C. or D.C.

Sonora-Model TSA-105
5 TUBE AC-DC SUPERHET
With Built-in "SONORASCOPE"

Imbued with some of the most modern developments in radio design, this exquisitely beautiful little Sonora model strikes a new note in small radio performance. Has a smart, modern molded cabinet with unusual louvre grille design. The beautiful soft lines give it an appealing beauty. Available in either soft ivory or in walnut. Tunes all standard Broadcast programs on 1720 to 535 k. c., including the 1712 Police Channel. Features include: Built-in "Sonorascope" . . . no aerial or ground needed; new 3-inch full-vision molded scale with molded tuning indicator control; P. M. Dynamic Speaker; Automatic Volume Control; 1¾ watts Output; Beam Power Output; new 150 Mil Tubes . . . no ballast or heating cord. Superhet circuit uses: 1—12A8GT, 1—12K7GT, 1—12Q7GT, 1—35L6GT, 1—35Z5. Operates on both AC and DC., 110 volts 40-60 cycles. Size: 8"x4¾"x5⅝". Ship. wt., 5¼ lbs.

As Low As

$8.07

No. **9905.** Walnut finish. List $12.95. Dlr's., ea. $8.65. Lots of 3, ea. $8.23, less 2%, net........ **8.07**

No. **9906.** Ivory finish. List $14.95. Dlr's., ea. $9.67. Lots of 3, ea. $9.29, less 2%, net........ **9.10**

Sonora-Model TV-48
4 TUBE AC-DC SUPERHET

As Low As

$6.27

Here is the new "Teeny-Weeny" sensation that sets the vogue for 1940. Truly remarkable in its powerful, dependable performance. The very spirit of tomorrow is captured in the sleek, modern lines of the superbly designed cabinet. It is plastic-molded in luxurious ivory or glistening rich onyx, with molded, full-vision type dial and dual-louvre grille. Tunes 1720 to 535 k. c. covering the entire standard American Broadcast band and the 1712 k. c. Police Channel. Features include: beam power 1½ Watts Output; P. M. Dynamic Speaker; attached 20-foot antenna hank; new 150 Mil Tubes . . . no ballast, no heater cord. The perfected Superhet circuit uses the following tubes: 1—12SA8GT, 1—12S17GT, 1—50L6GT, 1—35Z5GT. This beautiful and efficient model brings you outstanding qualities of eye and ear appeal with rare value that is certain of popularity. Operates on both AC and DC, 110 to 120 volts, 40 to 60 cycles. Sizes: 9"x5⅛"x4⅛".

No.	Color	List	Dlr's., ea.	3, ea.	Less 2%
9902	Walnut	$10.95	$6.90	$6.40	$6.27
9903	Ivory	12.95	7.50	6.85	6.71
9904	Onyx	14.95	9.27	8.91	8.73

In the 1940 Line, Sonora engineers and designers have achieved a degree of excellence in tonal quality as well as in decorative beauty, unequalled in radio annals. By virtue of its outstanding superiority, we can say with confidence that 1940 will see the name of Sonora carried to greater heights of popularity than ever.

Tuned to the Tempo of Tomorrow!

Sonora "CANDID PORTABLE"
MODEL KG-80

The TRULY PERSONAL RADIO

CARRIED LIKE A CANDID CAMERA

Sonora engineers have again blazed a glorious new trail—they have condensed Sonora's famous tonal beauty into even smaller bounds than has ever been attempted. In doing so, they have created a vast new market—the millions of outdoor loving Americans, who can now carry about their radio entertainment, the same as a handbag or a camera.

Entirely Self-Contained

No connections whatsoever, the Sonora "Candid Portable" is truly the personal radio. It can be carried about like a handbag or slung over the shoulder and worn like binoculars or a camera. On the golf course—on hikes—in canoes—on camping or fishing trips—the "Candid" makes a wonderful companion, keeping you completely entertained and apprised of world happenings. Amazingly economical, too, operating off of three ordinary flashlight cells and a 45 volt "B" Battery.

4-Tube Superhet Circuit

Uses 4 new miniature type tubes for powerful, sensitive tuning and surprising volume. Tunes from 535 to 1720 K. C. and the popular 1712 K. C. police channel. Tuning is greatly simplified by means of the full-vision dial. Has automatic volume control with surprisingly wide range of volume adjustment.

Built-in "Sonorascope"

No aerial or ground required. Equipped with a 4" P. M. Dynamic Speaker, the same as used in larger types radios. Reproduces with amazingly fine clarity and tonal beauty. Has a connection for plugging in head phones where privacy is desired. When used with head phones, speaker is automatically silenced.

Molded Cabinet of Sleek Beauty

In its sleek molded cabinet of black Durez and nickel plated trim, the "Candid" will certainly appeal to moderns. Only 8¼"x5"x4½" and weighing but 4½ lbs., it's the last word in compact beauty. Opens in half like a book for access to batteries.
No. 9954. List $17.95. Dlr's., ea. $11.51. Lots of 3, ea. $10.95, less 2%, net ....

10⁷³

Extra Battery Kit
No. 9955. Eveready Battery Kit for Above. 3—No. 950 flashlight cells and 1—No. 738—45 volt "B" Battery. List $1.80. Dlr's., per kit $1.25, less 2%, net **1.22**

No. 9640. Adjustable black leather shoulder strap for above. List 50c. Dlr's., ea. **33c**
No. 9641. Headphones for above. List $1.45. Dlr's., ea. 87c, less 2%, net.... **85c**

Complete with Tubes and Batteries.

The NAVIGATOR RADIO

It's Novel-It's Entertaining-It's Educational

A smart new innovation in radio in which distinctive decorative beauty and educational value are combined with superb radio performance. Just the thing for son's or daughter's study room or dad's den—or even in the living room, where space is at a premium.

Just Turn the World to Tune

Powered with a precision engineered superheterodyne circuit in which 5 tubes provide 7 tube performance, the Navigator achieves a remarkably high degree of tonal excellence. Clear, sharp tuning is accomplished by simply rotating the globe on its axis. The natural tone chamber provided by the hollow globe produces a rich, resonant tone comparable to that of a large, console type radio. Equipped with a 5" dynamic speaker. Tunes from 550 to 1720 K.C. and also the 1712 K.C. police channel. **Built-in Sonorascope**—no aerial or ground required. Switch and volume control in base.

Watch History Being Made

With the Navigator, you can follow the news of the world as it happens. The atlas is 10" in diameter and mounted within a beautifully finished brass mariner's wheel, graduated for quick reference, in longitude and latitude. Mounted on a 15" base of beautifully grained walnut. Stands 15½" high overall. Ship. Wt. 15 lbs.
No. 9956. List $29.95. Dlr's., ea. $17.97, less 2%, net ....

17⁶¹

7-TUBE Performance

Illustration showing position of 5-tube superhet chassis in Globe

A World of Radio Entertainment!

5 TUBES* . . . BUILT-IN AERIAL

5 Tubes* Include Power Rectifier Tube. A smart, bright, plastic Silvertone that's low in price and yet equipped with features to assure good performance for a radio of its size! Small enough (5⅞ inches high, 8 inches wide, and 5 inches deep) to use in the kitchen, bedroom, or office, yet beautiful enough in its rich, sparkling colors and classic, modernistic design to be a welcome addition to any room.

A newly designed superheterodyne circuit with 2-gang tuning condenser and two dual-purpose tubes for extra power; Automatic Volume Control that reduces fading and blasting—holds volume constant for continuous enjoyment. 5 to 1 dial ratio to permit more accurate tuning. Station coverage from 540 to 1700 Kilocycles. Large, rectangular dial with beautiful plastic fluorescent pointer. Excellent small set tone from good quality 4-inch dynamic speaker.

Built-in Loop Aerial for good reception of local stations. Where loop aerial reception is not satisfactory an outside aerial must be added to the connections provided. See Page 845 for Outside Aerials. IF YOU REQUIRE CONTINUOUS RECEPTION FROM STATIONS OVER 100 MILES DISTANT, WE RECOMMEND one of our larger Table Model sets on this or the preceding pages.

Far Better Than Most small radios at this price. Test and compare this little Silvertone on our **15-Day Trial** described on Page 839. R.C.A. and Hazeltine Licensed. For 100 to 125-volt, 25 to 60-cycle Alternating or Direct Current electricity. Shpg. wt., 6 lbs. **Mailable.**

57 K 07000—Walnut Finish Plastic.................$7.49
57 K 07002—Onyx Finish Plastic....................8.49

$7.49
Walnut Color Plastic

$8.49
Onyx Color As shown

AMERICA'S MOST POPULAR RADIO—5-TUBE*

5 Tubes* Including Rectifier Tube. 4 Insta-matic, Easy-to-Adjust Push Buttons.
Here is a success story, we believe no other radio can boast! A phenomenal record—more than 200,000 satisfied customers have purchased Silvertone's COMMENTATOR in a little over one year, thus according it the highest possible praise as AMERICA'S MOST POPULAR RADIO. It gives you more for your money than any similar radio we've seen. Compare the COMMENTATOR feature-for-feature with radios at $15.

A Triumph of Radio Design with its satin-smooth overall front and back design. You can see from the mirror view above that the back is as beautiful as the front (not the usual cut-off cardboard back you see on other radios). Lighted "Sunburst" dial of translucent plastic that curves rays of light and makes entire dial light up.

Selective and Sensitive Superheterodyne circuit eliminates the "overlapping" of stations so common in small radios. Station coverage from 540 to 1600 kilocycles. Automatic Volume Control maintains volume intensity at the proper level.

More Mellow Tone—the COMMENTATOR gives you full-bodied, clear tone you can really enjoy; the "acid test" of any small radio. This is achieved with heavy 4-inch dynamic speaker which provides a desirable balance between low and high notes.

More Features: Insta-matic push buttons of the newest "piano key" type, tune-in your favorite stations at the touch of a finger. Easy to set-up, just 5 minutes time and a simple screw driver adjustment. Call letters furnished. Built-in loop aerial provides for excellent reception in large cities, or near powerful stations. In areas where loop aerial reception is not satisfactory, an outside aerial must be attached. See Page 845 for Outside Aerials. For 100 to 125-volt, 25 to 60-cycle Alternating or Direct Current. Size is 6⅛ inches high, 10½ inches wide, and 5¾ inches deep. Shipping weight, 10 pounds. **Mailable.**

57 K 07004—Rich Burl Walnut Color Plastic with Gold Color Dial, Knobs, and Push Buttons...$10.25
57 K 07006—Ivory Colored Plastic with sparkling Red Dial, Knobs, and Push Buttons.........11.25
57 K 07008—Onyx Color Plastic with Gold Color Dial, Knobs, and Push Buttons..............11.25

Sears Offer Convenient Easy Terms.   See Inside Back Cover

$10.25
Walnut Color With Push Buttons

$9.25
Walnut Color Less Push Buttons

New Low Price Commentator Without Insta-matic Tuning

Identical in features and performance and appearance to the famous Commentators described above but without the 4 automatic station push buttons. Shipping weight, 10 pounds. **Mailable.**
57 K 07024—Rich Burl Walnut Color Plastic with Gold Color Dial and Knobs..................$9.25

COMPACT ELECTRIC RADIOS . . .

6-TUBE* AC and DC PLASTIC
With 1 or 2 Tuning Bands

6 Tubes* Including Rectifier Tube
The ultimate in a radio for those who prefer a neat, distinctive table model and yet want superior performance and commanding tone quality. Here's more radio for your money for on every count this large 6-tube plastic radio scores with latest features, performing ability and smart practical design.

Latest Superheterodyne Design for exceptional sensitivity—combined with a low noise converter that reduces background noises so common on small radios while tuning stations. Has many other features not ordinarily found on sets in this price range!

And for "Extras." Automatic Volume Control to keep volume uniform on all programs; a 5-inch full-response dynamic speaker that delivers natural, life-like tone. Dial coverage includes regular American stations from 540 to 1700 kilocycles including some police calls.

Built-in Loop Aerial provides excellent reception in metropolitan areas or near powerful stations and reduces static and interference. Where loop aerial is not satisfactory or for short wave reception an outside aerial must be attached to the connections provided. See Page 845 for Outside Aerial Kits. For 100 to 125-volt, 60-cycle Alternating or Direct Current (25-cycle models, 50c extra). Size, 9¾ in. high, 12⅞ in. wide, and 7¼ in. deep. See Page 839 for **15-Day Trial.**

57 K 07010—Walnut Color Plastic. Shpg. wt., 12 lbs. Cash..$12.95
Easy Payment Price ($2 Down, $2 a Month). Mailable.........14.00

$12.95
With American Tuning Band

$14.95
With Foreign and American Tuning Bands

Foreign-American 2-Band Model With Tone Control

For those who desire the added pleasure of shortwave reception, we recommend our 2-Band Model which brings in shortwave broadcasts from 5.6 to 16 megacycles. Same fine features of the radio described above except has four knobs and **2-Position Tone Control.**
57K07016—Rich Walnut Color Plastic. Shpg. wt., 1[...]
Easy Payment Price ($2 Down, $2 a Month)....

1941 Catalog Page

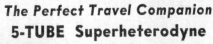

These are the new personal type radios that offer wider application of reproduced entertainment—whether at home, on an outing, going somewhere on a trip, aboard a train, bus or plane—anywhere with a set by your side you won't miss your favorite programs.

Imperial COMPANIONETTE
Personal PORTABLE RADIO

The Perfect Travel Companion

5-TUBE Superheterodyne

Remarkable lightweight set as small as a camera and as easy to carry. It is self contained with loop antenna and has on and off switch combined with door catch. Uses new 4523 Tube that reduces current consumption, making the set economical to operate.

3-WAY: AC — DC or BATTERIES.

Uses High Power Tubes

Marbled Effect Plastic Case with Simulated Tan Leather Trimming

Built-In Looptenna

Weighs only about 4 lbs.
Size 9x5x4½ in.

Plays anywhere anytime . . . on AC, on DC or on its self-contained batteries—the perfect radio to have along as a constant companion. Only 8¾ in. high, 4¾ in. wide and 4 in. deep, as easy to carry as a camera or a handbag—weight only 3½ lbs. **Uses the new 4523 tube** recently developed to decrease current consumption and heat. AC-DC cord plug-in may be disconnected when radio is not in use. This is the set to display—it catches the eye—the set to demonstrate—it charms the ear—**its performance is a revelation** . . . and the set to sell for volume business and greater profits. Powerful 5-tube superheterodyne circuit affords amazing reception—possible only since the development of the new tiny radio tubes. 5 tubes as follows: 1R5, 1T4, 1S5, 3S4, 4523. **Has a full tuning range of 540 to 1700 Kilocycles.** Automatic Volume Control, Easy Vision Tuning Dial, Self-Contained Looptenna. Uses two ordinary flashlight cells for "A" power supply, and No. XX45 "B" battery. Cleverly styled case resembling small camera of tan marble plastic and with pebble grain simulated leather trim, matching dial and speaker grille. Hinged door—"On and Off" switch is combined with door catch. Has easy-carrying handle. **Complete with Batteries.** LIST, 31.50.

No. G1R-D3782. Dealer's Price, each, 19.85.
Lots of 3, each, 18.75.
Less 2%, net.......................... **18³⁷**

Imperial
PORTABLE RADIO
Latest Personal Type

4-Tube Superheterodyne

Runs On Own Batteries

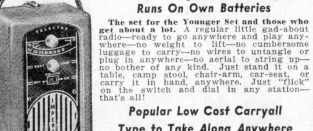

The set for the Younger Set and those who get about a lot. A regular little gad-about radio—ready to go anywhere and play anywhere—no weight to lift—no cumbersome luggage to carry—no wires to untangle or plug in anywhere—no aerial to string up—no bother of any kind. Just stand it on a table, camp stool, chair-arm, car-seat, or carry it in hand, anywhere. Just "flick" on the switch and dial in any station—that's all!

Popular Low Cost Carryall Type to Take Along Anywhere

No Antenna or Wires Needed

Leatherette Covered Camera Type Cabinet 9x5x4½ in. Only 3½ Lbs.

Small—only 5x4½ in. and 9 in. high and weighing only 3½ lbs.—this amazing 4-tube superheterodyne set packs a lot of power in its compact camera type case. With its handy strap handle on top it can be carried like a camera—taken anywhere, to college, to camp, on picnics, hikes, outings, sport events, trips, etc., to bring in music and entertainment for any spare moment. Has built-in loop antenna and dynamic speaker. **Complete with 4 tubes:** 1A7GT, 1N5GT, 1H5GT, 3Q5GT; and battery equipment: one 6½ volt "B" and two flashlight dry cells. Tan Leatherette covered Case. LIST, 14.50.
Complete with batteries.
No. 1R-T40. Dealer's, each, 9.27.
3, each, 8.68. Less 2%. **8⁵⁰**

Imperial 3-WAY RADIO
PORTABLE
WILL PLAY ANYWHERE

5-TUBE SUPERHETERODYNE

To achieve compactness two latest miniature type tubes are used. These pack tremendous power for their size insuring finest performance without affect on battery life. Automatic power shift prevents accidental tapping of battery when set is plugged in on house current.

PLAYS ON AC or DC or on Self-Contained BATTERIES

Automatic Power Shift

British Tan Cabinet with Slide-away Door Protective Lock Front

BUILT-IN ANTENNA

Outstanding value in a self-contained portable radio with long range station reaching power. Superheterodyne circuit with 5 tubes as follows: 1R5, 1T4, 1H5, 1T5 and 35Z5. Use of two latest type miniature tubes permits unusually compact design without sacrificing performance or battery life. Large built-in extra-sensitive loop antenna, beam power output gives full rich tone. **Automatic power shift saves battery of drain when set is plugged in on house current.** Full response dynamic speaker—artistic airplane dial. Compacted within De Luxe closed front cabinet with slide-in door, lock and key. **Covered in British Tan simulated leather** contrasted by pastel tan inside panel, stitched cowhide handle. Size 13½x8⅝x5⅞ in. Uses 1-6 volt "A" and 2-45 volt "B" batteries. (Approx. 200 hr. life.) **Complete with batteries.** LIST, 34.95.
No. 1R-HPT-51. Dealer's, each, 21.55. 3, each, 20.45. **18⁸⁶**
6, each, 19.25. Less 2%, net..............

Imperial 6-TUBE PORTABLE
3-WAY RADIO
Plays on AC-DC OR BATTERIES

Built-in Antenna

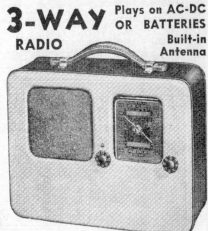

Plays anywhere from own self-contained batteries or from AC and DC house current without manual switching from one to the other. Built-in aerial. Highly selective set with a 6-tube superheterodyne circuit containing the following tubes: 1A7GT, 1N5GT, 1H5GT, 3Q5GT, 50L6GT, 3524GT. Full response dynamic speaker — Vernier airplane dial. **Tunes 540 to 1600 KC.** Automatic amplification Control. Uses two 4½ volt dry "A" batteries and two 45 volt "B" batteries in pack. In new two-tone grey and brown luggage style case. Size 13x10½x 5½ in. **Complete with batteries.** LIST, 34.95.
No. 1R-D3891. Dealer's, each, 21.95.
Lots of 3, each, 20.87. **20⁴⁵**
Less 2%, net..

5-Tube 3-Way Portable Radio

Plays from own self-contained batteries or from AC-DC house current. Has a five-tube superheterodyne circuit with the following tubes: 1A7GT, 1H7GT, 1NSGT, 1ABGT, 35Z5GT. Built-in aerial, dynamic speaker, lighted dial—**tunes standard broadcasts.** Uses same batteries as above. In grey and brown luggage case. Size 11½x8½x5½ in. **Complete with batteries.** LIST, 27.95.
No. 1R-D3831. Dealer's, each, 17.95.
Lots of 3, each, 16.95. **16⁶¹**
Less 2%, net..

STYLE LEADER 5-TUBE*

Battery, A.C. or D.C. Operation

5 Tubes* Including Power Rectifier Tube. Here's the best-looking and best performing portable radio in its price class! Smarter and more modern in design than many sets at $30 and fully equal in power and features to most sets at $25. You'll enjoy its superior tone and you'll be proud of its streamlined appearance.

$18⁹⁵ Cash

with Batteries
$2.50 DOWN

Operates 3 Ways—from 100 to 125-volt, 25 to 60-cycle A.C. electricity . . . from 100 to 125-volt D.C. electricity . . . and from 200 to 250-hour enclosed dry batteries. Use it anywhere outdoors . . . or when indoors where electricity is available, plug it in to save the batteries.

Improved Performance and sensitivity through new design. A special type of built-in aerial that is more efficient shields out noises and gives better-than-ever reception when you are near large broadcasting stations. In places where built-in aerial reception is unsatisfactory, an outside aerial must be attached. Economical superheterodyne circuit is easy on batteries. Tunes regular American stations from 540 to 1600 KC. Excellent tone quality from a new 5-inch dynamic speaker. See Page 839 for **15-Day Trial.**

Exclusive Design is trim and neat and beautifully proportioned. Its front grille is of sparkling **brushed-chromium** finish. Case is covered in sturdy washable airplane luggage fabric in a mixed **Brown** tweed effect with a contrasting dark brown simulated leather trim. Hinged cover protects dial and controls. Easy to carry it weighs only **17 lbs.** Size, 10½ inches high, 12¼ inches wide and 5½ inches deep. Shipped complete with one 57 K 5080 and two 57 K 5079 dry batteries. **Mailable.** Shpg. wt., 20 lbs.

57 K 07080—Portable Complete. Cash $18.95
Easy Payment Price ($2.50 Down, $3 a Month) 20.55

View above shows "saddle" cover removed for playing, and view at right shows cover in place for traveling.

5-TUBE* COMPACT PORTABLE

Battery, A.C. or D.C. Operation

5 Tubes* Including Power Rectifier Tube. Almost as small as the new camera-type portable radios, yet has all the tone and battery life advantages of larger portable sets. Use indoors on electricity and outdoors on batteries. Excellent tone—new alloy dynamic speaker.

$16⁹⁵ Cash

with Batteries
$2 DOWN

Weighs Only 11 Pounds, and is only 7½ in. high 11 in. wide and 5¼ in. deep. Case is covered in a warm brown pebble grain simulated leather. Tunes American broadcast stations from 540 to 1600 KC. Built-in loop aerial for reception when near larger broadcasting stations. Where loop aerial reception is unsatisfactory, attach an outside aerial.

When Indoors this portable operates from 100 to 125-volt. 25 to 60-cycle A.C. or D.C. electricity. When outdoors, it uses dry batteries that will play 175 to 200 hours. Complete with one 57 K 5086 and two 57 K 5072 batteries.

57 K 07076—Shpg. wt., 13 lbs. Mailable. Cash $16.95
Easy Payment Price ($2 down, $2 monthly) 18.40

DELUXE SILVERTONE 4-WAY PORTABLE

Complete With Detachable Radionet

6 tubes Including Rectifier Tube for Electric Operation. Silvertone presents the last word in portable radios— a new powerful model with an extra built-in aerial that is detachable for use in autos, trains, planes, boats— almost any place where ordinary portable sets fail. This amazing feature greatly increases the enjoyment you get from this portable—you can take it more places, and get better reception. You'd pay $8 to $10 more for it in other stores. Simply remove the special Radionet from the inside of the set, and attach it close to a window with the handy 2-way, hook-clip fasteners. In ordinary locations the regular built-in aerial picks up strong nearby broadcasting stations without even using the detachable Radionet.

$23⁹⁵ Cash

with Batteries
$3 DOWN

Smartly Styled like expensive luggage with a case covered in **blond simulated rawhide.** End boots and a removable cover are of contrasting **brown** artificial leather. Heavy cord stitching around end boots like fashionable luggage. Speaker grille and dial cover of beautiful exclusive design in modernistic brushed chromium finish. Easy to carry—weighs only **20 lbs.** and is 11½ inches high, 12¼ inches wide, and 6 inches deep.

Unusually Powerful and selective superheterodyne circuit with 6 tubes, PLUS a brand new achievement . . . *constant sensitivity.* This exclusive feature means you get peak performance throughout practically the entire life of the batteries—no dropping off of sensitivity when batteries grow weaker as in ordinary portables! 5-inch permanent magnet type dynamic speaker assures excellent tone quality with plenty of volume. Automatic Volume Control lessens fading of stations. Tunes American broadcast band from 540 to 1600 KC. Easy-tuning thumb wheels and slide-rule dial protected by hinged cover. See Page 839 for **15-Day Trial.**

4-Way Operation—from 100 to 125-volt, 25 to 60-cycle A.C. electricity . . . from 100 to 125-volt D.C. electricity . . . from powerful 250 to 300-hour enclosed dry batteries . . . and with separate detachable aerial for autos, trains, etc. Shipped complete with two 57 K 5085 and two 57 K 5090 batteries. Shpg. wt., 23 lbs. **Mailable.**

57 K 07084—Deluxe Dual Aerial Portable. Cash $23.95
Easy Payment Price ($3 down and $4 a month) 25.95

View above shows how detachable Radionet is placed in trains, autos and other places where reception conditions are difficult. When not in use, it fits in cabinet out of way—see the picnic scene below.

Radios on these pages shipped from Chicago, Philadelphia, Boston, Minneapolis, and Kansas City.
25-cycle models shipped from Chicago and Phila. only. Order from your nearest Mail Order House.

⊖ **PAGE 841** . . . . **RADIOS**

63

Modern Super-Power Table Model Radios

Imperial Smartly Designed De Luxe Superheterodyne
With All Advanced Scientific Features for Superior TONE

Powerful New Type Super - Sensitive Tubes Insure Best Possible Performance

Amazing reception over wide tuning range, 540 to 1700 KC—your popular programs, clear and free from intrusive noises—unexcelled tone quality, infinitely better than from many sets.

Exceptionally Fine Tuning and Perfect Station Separation
BEAM POWER OUTPUT — BUILT-IN PARASCOPE AERIAL

Scores of Other FEATURES

- Best Engineered 5-Tube Superheterodyne.
- Large 5 - Inch Electro-Dynamic Speaker.
- Automatic Volume Control.
- PARASCOPE, Enclosed Aerial.
- Extra Large Lighted Airplane Dial.
- Handsome WALNUT Cabinet.

Here is a set that will please those of your customers who want the rich distinctive beauty of a gorgeous cabinet in addition to the many internal features that are indispensable to superior reception. Equals sets costing twice as much—what a saving! Because in this Imperial you can enjoy the advantages of every exclusive development and effects of the greater drawing power of the built-in Parascope antenna. And, too, plus a **large 5-inch electro-dynamic speaker** that reproduces the entire musical register with fine fidelity, bringing into the home a purity of tone never dreamed of in a set of this class—actually recreating the stirring realism of the originating performance. Tone is kept at desired level by automatic volume control. **Covers tuning range from 540 to 1700 KC for standard American broadcasts.** Exceptionally large lighted airplane dial with 5½ to 1 cord drive, tunes stations speedily. The circuit includes the following 5 tubes: one each, 12SA7, 12SK7, 12SQ7, 50L6GT, 35Z5GT. Exquisite modern Walnut cabinet, size 13x7½x6½ in. For AC-DC power. LIST, 22.50.
No. 1RD429. Dealer's Price, each, **13.95**.
Lots of 3, each, **13.15**.
Less 2%, net.

Dealer's Price
12 88 NET EACH
In Lots of 3

12 88

This New Imperial IN MODERN CABINET OF FINEST WALNUT

Supremely beautiful cabinet with graceful curved front—modern richness of design—finest of American sliced walnut veneers, substantial on solid base. The vertical bar over grille adds to its charm. Hand rubbed high lustre finish. Size 13x7½x6½ in.

Imperial Streamlined 5 - Tube Superheterodyne
With Built-in Long Range Parascope Loop Aerial

Ultra Modern Plastic Cabinet With Striking Louvred Grille Front

Has everything constructive in well developed tone of extraordinary quality for a set in this class, recreating originating programs with a fidelity that is really amazing. Powerful and selective—gets stations over wide range, giving every advantage for the reception of favorite programs—tone is clear, entirely free from extraneous sound effects. Requires no external aerial unless still greater station range is desired—enclosed PARASCOPE antenna serves to get standard stations with the best entertainment. **Electro-Dynamic Speaker.** Tone volume is held to desired level automatically—no blasts nor fadeouts. Has economical beam power output with dual power tubes, 5 tubes in circuit as follows: 35Z5GT, 50L6GT, 12SQ7, 12SK7 and 12SA7. Lighted Slide rule dial and 2-Knob controls. Attractive modern plastic cabinet with grille front. Size 8⅞x6¼x4¼ in.
No. 1RW2501V. IVORY.
Dealer's, each, 11.25. Less 2%, net................... **1102**
No. 1RW12501. WALNUT.
Dealer's, each, 10.25.
Less 2%, net..................

10 04

Large Dynamic Speaker

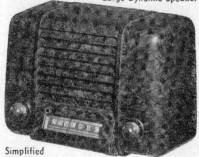

Simplified CONTROL

These Outstanding RADIO VALUES
Offer You Easier Selling and Greater PROFIT

This page offers you exceptional radio values—brand new sets, incorporating all improvements that make them desirable; and can easily be they can easily be basis of our low prices. dous purchases o concessions for us to save you some the orders you sell though, for limited these outstanding values will fast disappear.

ORDER NOW AND—
Suggest Radios for the APPROPRIATE GIFT

Imperial New Ideal Battery and 3-Way Farm Radios
These Are the Best Sets for Finest Broadcast Entertainment in Rurals!

Most versatile type radios offered on the market—most economical to operate—in eye-attracting cabinets of striking beauty, designed to the modern trend with curved fronts and made of genuine walnut woods.

Tone Quality Equal to Highest Priced Sets, Powerful P.M. Dynamic Speaker
- Finest Quality P.M. Dynamic Speaker and Tone.
- Full-Maintained Automatic Volume Control.
- Simplified Controls and Lighted Dial.

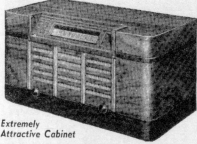

Wonderful New FARM RADIO
1,000 HOUR BATTERY LIFE
Imperial IN HANDSOME CABINET OF WALNUT WITH INLAYS

Singly on AB Pack or Separately on A and B Batteries

Unmatchable value! reflects the highest type of engineering ingenuity resulting in the most amazingly realistic tone reproduction possible. The superheterodyne circuit uses 4 low-drain 1.4-volt super-power tubes as follows: 1N5GT, 1A7GT, 1H5GT and 1Q5GT. Has full volume control—no blasts nor fadeaways. Tuning range 540 to 1630 KC. Beautiful inlaid walnut cabinet with oversize lighted vernier dial. Operates singly on AB pack or separately on A and B batteries for 1,000 hours' service.
Size 17⅝x8½x9¼ in. (Less batteries.) LIST, 33.00.
No. 1RW2402. Dealer's, each, **17.45**.
Less 2%, net.

17 10

USALITE 1,000 Hour Battery Pack for above.
90 V.B. and 1½ V.A. Size 16x4½x7⅛. LIST, 5.95.
No. 5RAB666. Dealer's, each, 4.16. Less 2%, net........ **408**

The Very Latest 3-WAY MODEL
Imperial DE LUXE CONVERTIBLE BATTERY POWER TO AC or DC

Ideal Set for Farm—Gives Service Ahead of Highline

Extremely Attractive Cabinet

Economical, uses only 4 tubes on battery—5 tubes on AC-DC. Automatic volume control—off-on indicator, P.M. dynamic speaker and automatic change over from farm radio to electric, requiring only addition of rectifier. Has tubes: 3Q5GT, 1H5GT, 1N5GT, 1A7GT. Slanted slide-rule dial. European type cabinet of walnut with stripe contrasts. Takes 1,000-hour battery pack. Size 16¾x8½x9¾ in. (Less Battery Pack and Rectifier.) LIST, 39.95.
No. 1RW2565. Dealer's, each, **20.50**.
Less 2%, net.

20 09

Usalite 1,000-Hour Battery Pack for Above.
90 V.B. and 1½ V.A. Size 16x4½x7⅛. LIST, 5.95.
No. 5RAB66. Dealer's, each, 4.16. Less 2%, net........ **408**
Rectifer Tube for Above.
No. 5R117Z6GT. Dealer's, each, 88c. Less 2%, net........ **86c**

New Pickwick Radios for Spring and Summer Sales

PICKWICK 5-Tube Plastics AC-DC

These new plastics in gorgeous colors of Ivory, Green, Red, Walnut and Black are the most beautiful of all small compact radios. Modernistic design, with curved horizontal louvres over speaker grill, is the predominant motive. The radio is a five-tube A.C.-D.C. set equipped with a powerful, full dynamic speaker providing a tone unusual in a set of this size; two band tuning; built-in aerial and a distinctive illuminated gold tuning dial in three colors.

- Standard Broadcasts
- Amateur and Police
- Modernistically Beautiful
- Dynamic Speaker
- Built-in Aerial

Operates on any 110-volt current, A.C. or D.C. Cabinet measures: 8¾x6¾x5. Tubes used are: 6D6, 6C6, 43, 25Z5, L49C. Shipping weight: 9½ lbs.

No. 623H43—Black. Retail $17.25.
Dealer, each **$9.95**
No. 623H44—Walnut. Retail $17.25.
Dealer, each **$9.95**

No. 623H45—White. Retail $19.50.
Dealer, each **$10.95**
No. 629H46—Green. Retail $19.50.
Dealer, each **$10.95**
No. 623H47—Red. Retail $19.50.
Dealer, each **$10.95**

PICKWICK 7-Tube Magic Eye AC-DC
SUPERHETERODYNE

A compact radio designed to give a small sized radio the power, tone, sensitivity and high grade performance of the Big Radio. Its advanced engineered 7-tube Superheterodyne provides a coast-to-coast tuning range with 2 band tuning from 75 to 555 meters, with an artistically designed dial in gold with dial readings in three colors. The Magic Eye permits accurate tuning almost automatically. New beam output tube provides tremendous undistorted volume. Full automatic volume control prevents fading on distant stations.

- Full-Sized Dynamic Speaker
- Long Distance Reception
- Magic Eye
- Beam Output Tube

The cabinet is of one piece rolled front and top in beautifully finished high grade, striped walnut with heavily receding fluting on both ends to give the cabinet an unusually massive appearance. Measures 13½x8x6. Shipping weight, 12½ lbs. Tubes: 6D6, 6G5, 6A7, 75, 25L6, 25Z5, L49C. Operates on any 110-volt current, A.C. or D.C.
No. 623H48—Retail $26.65.
Dealer, each **$16.00**

PICKWICK 8-Tube 3-Band AC-DC
MAGIC EYE SUPER

1941 Catalog Page

...t that has everything. Giant Gold 6-inch tuning dial for ease of three band tuning with separate colors ... band; new beam output tube for tremendous undistorted volume, automatic volume control; tone ... dynamic speaker; new coil assembly that assures European reception; and a cabinet that is a master-... h and craftsmanship; AND Magic Eye—the radio engineering invention that takes guesswork out of ... tuning.

- Foreign Reception
- Beam Output Tube
- Magic Eye
- Giant Gold Dial
- Tone Control

The cabinet is wrought of matched walnut and maple front rubbed to a piano finish with highly effective fluting on top. Cabinet measures 19½x11x8. Shipping weight, 19½ lbs. Console model, not illustrated, 37x20x10. Console model, 43 lbs. Tubes employed are: 6A7, 6D6, 76, 6K5, 25L6, 25Z5, L40S, 6G5. Operates on any 110-volt current, A.C. or D.C.
No. 623H49—Table Model. Retail $38.25.
Dealer, each **$22.90**
No. 623H50—Console Model. Retail $50.00.
Dealer, each **$30.00**

PICKWICK 7-Tube World-Wide AC
MAGIC EYE SUPER

- Foreign Reception
- Magic Eye
- Tone Control
- 3-Gang Condenser
- Extra Stage R.F.
- Exclusive Coil Assembly
- Giant Gold Dial
- Automatic Volume Control
- Super-Sensitivity

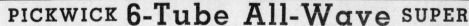

The Pickwick 7-Tube Magic Eye Superheterodyne is the apex of radio engineering, of precision workmanship and advanced styling. From a special highly developed coil assembly (built as a separate fixed unit, pre-determined as to its electrical efficiency before placed in the receiver and incapable of alteration), coupled with a three-gang condenser employing an extra stage of R. F., is evolved a state of sensitivity and selectivity that will cause shocked surprise to the most critical and technical radio enthusiast. Foreign reception is absolutely guaranteed. Tune Foreign Broadcasts from 17 to 52 meters; Foreign, Short Wave, Ship to Shore, and Amateur from 52 to 175 meters; Coast-to-Coast Domestic Broadcast from 175 to 555 meters.
The Giant 6-Inch Gold Dial, calibrated in three bands with separate colors for each band, used with the Magic Eye makes for tuning both accurate and delightfully simple. Powerful dynamic speaker, tone control and A.V.C. give to this model everything to be desired. The Cabinet is of beautifully matched walnut and mahogany, hand rubbed to a piano finish with vertical flutes on one end and a rounded one-piece front and side at the other end of cabinet. Measures 20x11x9. Weight 23 lbs. Tubes: 6K7, 6A8, 6K7, 6Q7, 6F6, 5Y3, 6G5.

No. 623H51—Retail $41.25.
Dealer price, each **$24.75**
No. 623H52—Console Model, not illustrated. Measures 37x10x10. Shipping weight, 48 lbs. Retail $61.25.
Dealers, each **$36.70**

PICKWICK 6-Tube All-Wave SUPER

The Pickwick 6 is a 110-volt 50-60-cycle A.C. quality receiver, precision made and expertly engineered to a degree of sensitivity and selectivity uncommon to the best of radios. Its three gang condenser with an extra stage of R.F. permits 5-10 kilocycle selective tuning and a state of sensitivity that brings in two to three stations between every point on the dial clear across he band. Large 6-inch gold tuning dial with individual colors for each of the three wave bands makes for beauty and ease of tuning. Foreign reception guaranteed. Powerful dynamic speaker makes for tonal delight. Tunes from 17 to 555 meters. Incorporates automatic volume control and tone control.
The Cabinet is a beautifully piano finished upright model of high-grade figured butt walnut with fully rounded sides and one-piece front and top. Measures 13x17x9½ ins. Tubes: 6K7, 6A8, 6K7, 6Q7, 6F6, 5Y3. Shipping weight 23 lbs.
No. 623H53—Retail $37.50.
Dealers, each **$22.50**

6-TUBE 6-VOLT BATTERY SUPER

The Pickwick 6-tube, 6-volt battery operated superheterodyne is a three band receiver covering 17-555 meters. Requires only one 6-volt storage battery. Absolutely humless and low drain. Has an output, tone, sensitivity and volume of the best A.C. receivers. A De Luxe Radio. Employs Permanent Magnet speaker. Tubes: 6D8, 6S7, three 6L5, 19. Shipping weight 25 lbs.
No. 623H54—(Less battery.) Retail $46.60.
Dealers, each **$27.95**

6-TUBE 32-VOLT SUPER

The Pickwick 6-tube, 32-volt superheterodyne operates directly from 32-volt farm power plants with no additional attachments or resistors. Tunes three bands from 17-555 meters. Dynamic speaker. Tubes: 6A7, 6D6, 75, 76, 48. Shipping weight 19½ lbs.
No. 623H55—Retail $39.90.
Dealers, each **$23.95**

PLAY ANYWHERE—Take Along on Outings, On Trips—To Homes of Friends—Just as Easy To Carry as a Lightweight Overnight Case—Just as Compact—Just as Handsome

Imperial De Luxe 3-WAY 6-TUBE SUPER IMPROVED PORTABLE

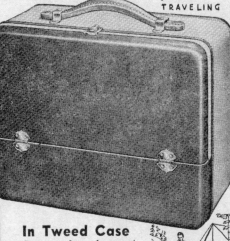

TRAVELING

3-WAY

PLAYS AC Current in the Home 1

PLAYS DC Current on Trains in office or Store 2

PLAYS from its own BATTERIES Anywhere 3

Offers the finest in entertainment any-where—no matter where, aboard the train, on steamship, in auto, camp, seashore and in the home—to amuse, enlighten and en-tertain. By means of its powerful circuit this set can be played in distant places where the ordinary portable will not op-erate. Tunes standard broadcasts. Finest triple purpose superheterodyne circuit with 6 powerful tubes, one each as fol-lows: 1A7GT, 1M5GT, 1H5GT, 3Q5GT, 35Z4GT and 50L6GT. Battery drain is sur-prisingly low. **Special change-over switch prevents drain on battery when line cord plug is used to operate from house cur-rent.** Has double power output on AC-DC, increasing range and is highly selective. True-tone super-powered P.M. Dynamic speaker and automatic amplification con-trol. New Par-a-Scope aerial with wide reach—needs no external antenna nor ground. Uses two 4½ volt dry "A" and two 45 volt "B" batteries. **Shock proofed in sturdy portable luggage type case to stand travel and vacation handling,** two-tone Parakoid covered with Sun-Tan leather handle and piped trim. Has ship design grille, hinged front panel on deco-rative hinges and with fastener. Size 12½x10½x6 in. 3-way AC-DC and battery, complete with batteries. Wt. 12 lbs. LIST, 39.95.
No. IR-D3893. Dealer's, each, 25.95. 3, each, 24.45. Less 2%, net.... **23⁹⁵**

Efficient Built-In PARASCOPE Aerial

- Tunes Standard Broad-cast Band
- Built-in Aerial—no antenna—no ground needed
- Super-Powered P.M. Dynamic Speaker
- Automatic Change-over Switch Control

In Tweed Case
Same as above, in smart grey and brown tweed fabric covered case. Size 12½x9½x6 in. **Complete with batteries.** LIST, 37.50. No. IR-D389. Dealer's, each, 24.25. Lots of 3, ea., 23.25. Less 2%, net. **22⁷⁸**

CAMPING

Imperial New Automatic Power-Shift 3-Way Superhet

PORTABLE **6** **Latest Type Super-Sensitive Tubes** **Extremely Low Battery Drain**
WILL PLAY ANYWHERE—Extra Long Range Special Built-In Loop Antenna

Streamlined

Dealer's Price Each, 25.25, 6, ea. **23⁷⁵ Net**

- Full Response 5 in. P.M. Dynamic Speaker
- Gets Broadcasts on Full Standard Band
- Automatic Power-Shift Safeguards Battery
- Streamlined Portable Type Luggage Case

Extra powerful 6-tube superheterodyne with greater coverage of broadcast stations, combining excellence of tone with conve-nience of 3-way power, in a lightweight luggage type portable case. Operates on AC or DC or self-contained batteries with batter-ies giving approximately 200 hours' service. Uses two 4½ volt "A" and two 45 volt "B" batteries. Automatic power shift from bat-teries to AC or DC operation prevents accidental draining of bat-teries when used on house current. Latest GIANT power minia-ture tubes as follows: 2-1T4, 1-35Z5GT, 1-1H5, 1-R5 and 13Q5. These tubes allow compactness without sacrifice of performance or battery life. Large 5-in. heavy permanent magnet speaker, extra sensitive built-in loop antenna and all other engineered units of highest quality. Sturdily built streamlined shock proof case, has cut-away drop-down removable front cover with solid brass English lock and key. Rich grained tan simulated leather covered contrasted with lighter tan inside panel—strong durable stitched cowhide handle. **Will withstand travel and rough hand-ling.** Size 14¼x8⅜x6½ in. Complete with battery pack. LIST, 39.95.
No. IR-HPQ-61. Dealer's Price, each, 25.25. 3, each, 24.50. 6, each, 24.25. Less 2%, net... **23⁷⁵**

Imperial 5-TUBE 3-WAY PORTABLE For AC-DC or 1½ V. Battery
Compact in Easy-carrying Tweed Covered Case

Operates from batteries or 110 volt AC or DC. Requires no aerial or ground—complete with built-in antenna. Has P.M. dynamic speaker, and easy-to-read dial. **Reception over the entire broadcast band from 540 to 1550 KC.** Stations come in clear and tuneful. Powerful super-heterodyne circuit with 5 full working high fidelity tubes. Luggage type tweed covered case. Com-plete with tubes and 250-300 hour rated battery pack. LIST, 26.95.
No. IR221. Dealer's, each, 17.85. 3, ea., 16.95. 6, ea., 16.25. Less 2%, net. **15⁹²**

Same as above, except with hinged front cover. LIST, 29.95.
No. IR507. Dealer's, each, 18.95. 3, each, 18.10. 6, each, 17.34. Less 2%, net.... **16⁹⁹**

Imperial RANGER SUPERHETERODYNE 1½ VOLT POWER A POPULAR PORTABLE
PRICE CUT TO CLOSE OUT LIMITED QUANTITY

Priced low enough for everyone—affords reception fine enough for anyone. Range 550-1750 KC (560-172 meters). **Standard broad-casts and police calls.** Full Automatic Volume Control. Improved new type lattice wound built-in antenna—needs no external aerial or ground. 5½ in. permanic speaker. Wide Vision dial with unique "on-off" warning signal. Ex-tremely low battery drain. Set and battery compartment in airplane luggage type case, covered with attractive water-proof alligator grain fabric. Size 11⅝ in. wide, 6 in. deep, 9⅜ in. high. 4 tubes: 1A7G, 1H5G, 1N5G, 1C5G. Less Bat-teries. 10 lbs. LIST, 19.95.
No. IR554BA. While they last, Each, 8.75. 3, each, 8.45. 6, each, 8.15. Less 2%, net....... **7⁹⁸**
No. 2RU2434. Power Kit for Above. Contains 1½ volt "A" and 90 volt "B" power. LIST, 4.00. Wholesale, per kit, 2.40. Less 2%, net............. **2³⁵**

66

GENERAL ⟨GE⟩ ELECTRIC

GoldenTone *Superheterodynes*

Every Proven Basic Principle Plus Latest Engineering Features for Best Radio Reception
Magnificent Sets Backed By This Famous Name—All with G. E. Beam-A-Scope Aerial

Beautiful Cabinet of Genuine Walnut
Built by master-craftsmen from exquisitely grained **genuine walnut.** Front corners curved — contrasting dark band around base — new slide rule dial. Size 13x9¼x6¾ in.

GENERAL ⟨GE⟩ ELECTRIC
Golden Tone
6 TUBE SUPER
Master DeLuxe Model
Equals Efficiency of
8 TUBE SET
G. E. Beam-A-Scope Aerial

- *6 Tubes with Efficiency of Regular 8-Tube Set.*
- *BEAM-A-SCOPE Aerial requires no External Antenna nor Ground.*
- *Standard and Police Band.*
- *Large 6-inch Dynamic Speaker.*
- *Walnut Cabinet.*
- *Illuminated Slide-Rule Dial.*

All of the quality and advanced features that has made G.E. radios the choice of thousands. Super sensitive and highly selective—picks up broadcasts over a wide range tuning the entire broadcast band, covering **540 to 1720 Kilocycles** with intermediate frequencies of **455 K.C.** Has G.E. Beam-A-Scope Antenna system—needs no ground nor external aerial. Large 6-in. dynamic speaker. The superheterodyne circuit has 6 matched G.E. pretested tubes as follows: 1—12SA7, 1—12SK7, 2—12SQ7, 1—35L6GT and 35Z5GT. Handsome walnut cabinet with new horizontal type illuminated slide rule dial; calibrated for easily locating station. AC or DC operation.
No. **IRGE-LCP609.** Dealer's Price, each, 16.75.
3, each, 15.96.
Less 2%, net................. **15⁶⁴**

G. E. Golden Tone 5-TUBE Superhet

Precision Built GENERAL ⟨GE⟩ ELECTRIC **2-Band Model**

More for the Money — Extremely Powerful with Performing Efficiency of A 7-Tube Radio Set—Has G. E. Beam-A-Scope Aerial

- *Standard Broadcasts and Short Wave.*
- *Large Illuminated Dial with Pointer.*

Highly sensitive 5-tube superheterodyne with G.E. matched pre-tested tubes as follows: 1—12SA7, 1—12SK7, 1—12SQ7, 1—50L6GT and 1—35Z5GT. Built in loop G.E. Beam-A-Scope antenna. Two Bands; standard broadcasts are reached over 535 to 1630 K.C.—shortwave police, amateur and aircraft 2.8 to 6.58 megacycles. Sound reproduction is faithfully developed and then amplified clearly by the full size 5-in. P.M. speaker. Automatic volume control maintains volume intensity and prevents blasting and fadeouts. Modern streamlined ivory bakelite cabinet with illuminated dial, artistically calibrated and with pointer. Size 13½x8x7 in. Operates on AC-DC. 24.00.
No. **IRGE-596SW.** Each, 13.75.
3, each, 13.20. Less 2%, net...... **12⁹⁴**

Same, in slightly different design cabinet. One Band—regular broadcasts from 540 to 1670 KC. 20.95.
No. **IRGPSW596.** Each, 13.07.
3, each, 12.15. Less 2%, net... **11⁹⁰**

G. E. Golden Tone 4-TUBE Superhet

Today's Outstanding Radio Value

GENERAL ⟨GE⟩ ELECTRIC

- *4-Tube Super with 5-Tube Efficiency.*
- *Standard Broadcasts and Police Calls.*

Ideal extra radio—in spite of tiny size has tremendous receptive range and power. Neat appearing walnut finish bakelite cabinet 7½x4¼x5 inches. Has superior 4-tube superheterodyne circuit developing power that equals most 5 tube sets. Contains the following G.E. pre-tested tubes: 12SA7, 12SQ7, 35Z5GT and 50L6GT. Volume is regulated automatically and controlled to desired intensity. Large dynamic speaker gives clear tone. Sensitive and selective. Gets standard broadcasts over range of 540-1720 KC, also police calls. Operates on AC-DC power sources.

No. **IRGE-LC401.** Dealer's Price, each, 9.75. 3, each, 9.15. Less 2%, net.... **8⁹⁷**

TRADITIONAL QUALITY

Offers all of the excellence and prestige of the familiar G.E. name—everything for perfect tone reproduction, created by modern science—superior performing instruments, attractive in the elegance of new styled cabinets. **Better and clearer reception with wider range of pickup** for more extensive choice of broadcast entertainment—effectively developed by the **greater power in these wonder radios,** with the internal G.E. Beam-A-Scope.

GENERAL ⟨GE⟩ ELECTRIC
GoldenTone
5-TUBE Superhet
With G. E. Beam-A-Scope Aerial
Tuning Frequency Range 540 to 1720 KC

The Ultimate in Distinctive Sets
SMART BAKELITE CABINET
—AND THESE FEATURES

- *Superheterodyne Circuit.*
- *Automatic Volume Control.*
- *Full Range Volume Selector.*
- *Lighted Slide Rule Dial.*

AT NEW LOW PRICE

Everything desired in fine broadcast entertainment becomes available to your customer when tuning in on this set. Compactly built table model in the new style—handsome ivory bakelite cabinet with red knobs and attractive red illuminated slide rule dial with white calibrations. The superheterodyne circuit is equipped with 5 of the latest G.E. matched pre-tested tubes consisting of one each as follows: 12SK7, 12SA7, 12SQ7, 35Z5GT and 50LGGT. Having a G.E. Beam-A-Scope, no external aerial nor ground is required. Volume is maintained by **automatic control**—keeps intensity regulated, avoiding "blasts" and fadeouts. Powerful P.M. Dynamic Speaker. Extreme sensitivity and fine selectivity—tunes American Standard Stations covering the broadcast band of frequencies from 540 to 1720 kilocycles with intermediate frequency of 455 K.C. Cabinet size 9¼x6x5½ in. Operates on AC-DC power source.
No. **IRGE-OLCP503.** Dealer's Price, each, 12.64.
3, each, 11.80. Less 2%, net.... **11⁵⁶**

Lots of 3, each,
11⁵⁶ Net

STEWART-WARNER

New Giant Power 6-Tube *Superheterodyne*

AMAZING 8 TUBE PERFORMANCE

NEWEST 1942 MODEL

Dynamic Speaker with Built-In Acoustical System Built-In LOOP AERIAL and these other STEWART-WARNER FEATURES

A real "hit" has been scored with this famous name set—a "hit" in style—a "hit" for performance—a "hit" from every angle, because in STEWART-WARNER you can be sure of getting quality. No other 6-tube superheterodyne today brings as many practical features, such excellence of tone, ease of tuning and such beauty in its superb modern styled cabinet, at so low a price. It's an amazing set, packing exceptional power, **giving 8-tube performance.**

KEEN STATION FINDING

Extra Power — Extra Sensitive

With this outstanding radio your customer can get the widest range of broadcast entertainment and finest tone reproduction of which any set is capable, regardless of price.

A Special Value at This Low Price

13⁶⁷ NET Each, in Lots of 6

Here you can get more for your money when buying this STEWART - WARNER 6-tube set because its super-heterodyne circuit includes extra-value features. Preserves proper balance between bass and treble.

Offers Big Chance for Extra Profit

This far-reaching set, with its 7 tuned circuits, obtains programs from a wide group of stations, covering a tuning range from 540 to 1720 KC., easily brought in and tuned at peak for perfect reception. Has two-position tone control, built-in LOOP antenna, large Deluxe 6 in. P.M. dynamic speaker—resistance-coupled beam power audio system. Simplified three knob control system and large lighted airplane dial with illuminated Lucite tip pointer. Chassis contains 6 high power tubes as follows: 12Q8GT, 12SQ7, 35L6GT, 35Z5GT, and 2-12SK7. Distinctive upright cabinet of molded plastic in Walnut finish **with acoustically correct built-in special design sound chamber.** Size 13x12¾x7¾ in. For AC-DC Operation.
No. G1RSW206. LIST, 29.95.
Dealer's, each, 15.75. 3, ea., 14.95.
6, each, 13.95. Less 2%, net............ **13⁶⁷**

Don't Wait SUPPLY Is LIMITED

Though we prepared for exceptional demand with a large quantity of these sets, the value is so sensational that even this apparently greater supply certainly can not last long. If you would benefit by the advantage our extremely low price makes available, order immediately as many sets as possible, even stretching a point to get a larger share of these fine radios. Buy early to avoid disappointment.

6-Inch DE LUXE Speaker Improved type oversize permeability alloy Dynamic Speaker. Specially constructed housing for perfect cone alignment.

Large Lighted Airplane Dial Exclusive new design with double pointer. Covers complete tuning band—fully calibrated. Simple controls.

2-Position TONE CONTROL Tone control gives a wide range of bass and treble as desired. Automatic volume control maintains volume intensity to proper sound level.

7 Tuned Circuits Assure better selectivity — better tone, smoother, more accurate dialing. Insures treble and bass notes in full tone.

Sensational DETROLA

Combination Radio-Clock
6-Tube 2-BAND SUPERHETERODYNE Foreign and American Broadcasts

NEW MODEL

Short Wave Reception Includes:

FOREIGN, AVIATION, AMATEUR and Some POLICE—Two Principal Bands

Standard American Broadcasts—Short Wave with 5 Subsidiary Bands

Note these Features

- 6-Tube AC-DC Superhet
- Two Principal Bands: Standard and Shortwave with 5 subsidiary bands
- Large 5 in. Electro Dynamic Speaker
- Automatic Volume Control
- Large Lighted Airplane Dial
- Reliable Electric Clock
- Exquisite Walnut Cabinet in Pleasing Colonial Style

IMPORTANT NOTICE

1941 Catalog Page

Built-in DETROSCOPE Aerial-Ground

In DETROLA is perfected the only practical built-in aerial developed in years. Named appropriately, the DETROSCOPE, this aerial of special circuit and windings, insures perfect reception anywhere and in areas where interference from signs, motors, elevators, etc., usually cause ordinary sets to perform in an unsatisfactory manner.

A Dual Purpose Instrument in a Famous Make

Aside from its unique styling which in itself is strongly appealing, this set combines two distinctly advantageous features—highest quality radio and the dependability of electric clock time keeping. The radio chassis contains fine AC-DC Superheterodyne circuit. 6 tubes, one a dynamic coupled output tube; two principal bands, one standard for American broadcasts, the other for short wave with five subsidiary bands—foreign, amateur, police, etc. **Covers 540 to 1720 Kilocycles (5.5 to 18.5 MC) and in the 5 Subsidiary Short Wave Bands 16 to 14 meters.** 5 in. electro dynamic Speaker, automatic volume control and other features of the best sets —large lighted airplane type dial with calibrations and band markings. Clock dial with sweep second hand. Clock operates on AC only.

Built-in DETROSCOPE aerial, requires no external aerial or ground—connection is provided for shortwave antenna.

Catalog Number G1RD3281C
Regular List, 31.45.
Dealer's Price, each, 15.95.
3, ea., 14.85.
Less 2%, net **14⁵⁵**

Colonial Style Walnut Cabinet

This DETROLA combination Radio and electric clock is contained in a Colonial style cabinet; proportioned from choice grained woods. Has butt walnut face panel bordered by fine marquetry and figured straight grain walnut veneers. Ideal for mantel or table in any home. Size 12½x10½x7½ in.

Imperial Decorative Utility Radios
New Distinctive Designs for Every Purpose
Fast Selling—Unique and Practical—FINEST PERFORMANCE

Here is a group of today's most decorative radio sets, offering at low cost excellent radio reception, unique beauty, and extra utility.

BRAND NEW NOVELTY
Just Introduced
Imperial De Luxe
• • • •
BABY GRAND
5-Tube Radio

Authentic Baby Grand Piano Design, Midget Size, Built with the Thoroughness of Its Life-Size Prototype

An exact reproduction of an eighty-eight note, eight octave piano keyboard, with keyboard made of Molded Tenite. The pedals are brass. Genuine piano hinges are used for the cover and the music rack.

Open view showing piano cover raised permitting free access to radio set.

Powerful 5-Tube Superheterodyne
Covers STANDARD AMERICAN Broadcasts and POLICE CALLS
7-TUBE PERFORMANCE

Covers full tuning range of 540 to 1720 Kilocycles.
- Built-In Synchrotenna
- Beam Power Output

Choice of Cabinets:
- Blond Prima Vera
- Genuine Walnut

Closed view showing piano cover down. In this position cannot be recognized as a radio set.

Never before a radio novelty like this! Actually —except for size—it cannot be distinguished from a real piano, the reproduction is so perfect. A beautiful ornament—a remarkable radio. Exquisite cabinet of either genuine Blonde Prima Vera wood or genuine Walnut with inlaid 2-tone panel. Keyboard of molded Tenite—exact reproduction of an 88-note, 8-octave piano. Both cover and music rack have genuine piano hinges. Pedals are finished in glistening brass. Entire case is hand rubbed to a finish of highest piano luster. Cover is so made that it becomes a magnificent sounding board, producing an astonishingly rich tonal effect. Superlative 5 tube superhet radio gives 7-tube performance. Tubes: 1 each 12A8, 12Q7, 12K7, 5016GT, 35Z5GT. Has 5 in. P.M. aligned dynamic speaker. Size: open, 11x10⅝x10 in. high; closed, 8 in. high. For AC or DC. LIST, 34.50.
No. IRG6534B. Blond Prima Vera.
No. IRG7534W. Genuine Walnut.
Each, 22.60. 3, ea., 21.33.
6, each, 20.35. Less 2%, net........ **19⁹⁴**

Imperial
MELODY CRUISER
Decorative Ship Model
With Famous
ARVIN Superhet RADIO

- Superheterodyne Circuit
- P.M. Dynamic Speaker
- New Heat Resisting Tubes

Sail Ho! When this full-rigged vessel sails into your customer's line of vision, it's "all hands on deck" to SEE it, HEAR it—and BUY it. A great seller everywhere—in big stores, small stores, department stores, jewelry stores. Radio is the latest model 4-tube Arvin superheterodyne, affording wonderful mellow tone and full volume. Tubes: 12SA7, 12SQ7, 35Z5GT, 50L6GT. Permanent magnet dynamic speaker. Hull is walnut finished, hand rubbed and highly polished, with unique anchor design speaker grille. Ship's rigging and sails act as aerial. Spars, sails, halyards and fittings are glistening chrome. Height, 16½ in., length, 19½ in. For AC or DC. LIST, 19.50.
No. IRMC029. Dealer's Price, each, 13.07.
Less 2%, net......... **12⁸⁰**

Imperial PORTO BARADIO
New Selling
Built-In Radio and Server Combined

Powerful 5-Tube STEWART WARNER Superhet Chassis

- 7-tube performance on 5 tubes
- Built-in loop antenna
- P.M. Dynamic speaker

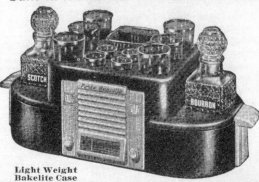

Light Weight Bakelite Case

Be the first in your community to show this most novel, most useful double-utility Porto Baradio. It will break sales records. Combines the PortoBar—immensely popular since its very introduction—with the famous Stewart-Warner 5-tube superheterodyne chassis that has achieved an unparalleled reputation. Includes 2 glass decanters, 6 highball glasses, 4 jiggers, glass ice tray, 6 glass mixers and ice prong—a complete 21-piece bar. Chassis is protected from liquid damage or reception interference. 5 tubes: 12SA7, 12SK7, 12SQ7, 50L6GT, 35Z5GT. Includes 10 ft. detachable cord with separable plug. For AC or DC. LIST, 31.95.
No. IRP900. Walnut with Ivory Grille, Ivory Handles.
No. IRP800. Black with Ivory Grille, Ivory Handles.
Dealer's Price, each, 19.16.
Less 2%, net..... **18⁷⁸**

Ivory Porto Baradio with Gold Grille and Gold Handles. LIST, 33.95.
No. IRP876. Dealer's Price, each, 20.36.
Less 2%, net........ **19⁹⁵**

Imperial TROPHY
Bowling Ball
5-Tube Superhet Radio

Designed for America's sports-minded millions, has universal appeal. 5-tube superhet chassis in black plastic "Bowling Ball" housing, realistic even to thumb-and-finger holes. Ivory plastic base, polished gold metal bands. 11 in. high, base diam. 7 in., ball 8 in. Chassis has built-in loop aerial, P. M. dynamic speaker, automatic volume control—and other features of the best radios. Covers standard broadcast 1730-540K.C. Provision for outside aerial. Tubes: 1 each 35Z5GT, 50L6GT, 12SQ7, and 12SA7. For 110 volt AC or DC. LIST, 20.95.
No. IRBB7936.
Each, 13.60.
3, each, 12.57.
Less 2%, net..... **12³²**

TROPHY Baseball
5-Tube Superhet Radio

Incorporates the same chassis as in the radio above. Giant white baseball, same dimensions as Bowling Ball Radio. Red "stitching," black printing and gold name plate. Black plastic base, polished gold metal bands. Overall height 10 in. For 110 Volt AC or DC. LIST, 18.95.
No. IRB9426.
Each, 12.30.
3, each, 11.37.
Less 2%, net.... **11¹⁴**

ELECTRIC CONSOLE RADIOS

1941 Catalog Page

FOUR-STAR FEATURE

★ **Outstanding Value!** Out-classed Seven Other Makes of Radios in Shopping Test.

★ **Superior Performance!** Latest type 3-gang condenser for sharper, clearer reception.

★ **12 Tonal Ranges!** Treble and Bass Boost Push Buttons PLUS 3-Point Tone Selector.

★ **Lasting Beauty!** Stylish Chest-on-chest Cabinet in hand-rubbed durable finish.

$49.95 Cash
$5 DOWN

8-Tube Superheterodyne Including Power Rectifier Tube
Awarded the Four-Star Seal of Approval because it offers so much in responsive wide range power and glorious tonal fidelity. It was built to the highest standards of technical perfection . . . performance-tested and laboratory-proved in order to win Sears famous 4-Star Approval. And best of all . . . this grand Silvertone is offered to you at about $30 less than other nationally advertised models of comparable quality. There's no padding in Sears prices to include high trade-in allowances . . . you can keep your old set and still save money.

Two S-p-r-e-a-d Bands that separate crowded short wave foreign stations. Simple, easy tuning of foreign programs on 9.4 to 9.8 MC. and 11.0 to 12.0 MC. ranges which cover European and South American stations. Two other tuning bands—540 to 1700 KC. for American broadcast stations, and 6 to 18 MC. for other short wave stations.

Eight Push Buttons—five are for instantly tuning your favorite stations at a finger touch. These can be easily set up in a few minutes' time with a screwdriver. Call letter tabs furnished. There are also **Two Tone Booster Buttons**, one for Treble and one for Bass note emphasis—see the complete story on this remarkable feature on the first radio page. Then there is an **Add-A-Unit** button for switching in a Record Player, or a Television or Frequency Modulation Converter at any time in future.

12 Tonal Variations at a Finger Touch

Big, 10-Inch Dynamic Speaker handles all tonal ranges and volume with amazing fidelity. Automatic Bass Compensation maintains the natural tone balance, strengthens the deeper tones that are so often lost in radio reception. A 3-position Tone Control knob operated in conjunction with the Bass Boost and Treble Boost push buttons gives you a total of 12 possible tone variations to suit any music or mood. Steady, smooth volume due to Automatic Volume Control.

8-Tube Superheterodyne Circuit with three-gang condenser, Low Noise Converter Circuit and expensive Push-Pull audio system give you extra sensitive and selective precision tuning and smooth, effortless volume of sound.

Built-in Semi-Rotating Loop Aerial gives you excellent reception in areas near broadcasting stations. Just plug-in the set and play. Where built-in loop aerial reception is not satisfactory, and for Foreign programs, an outside aerial must be added to the connections provided. See outside aerial kits on Page 845.

Lovely Rubbed Finish Cabinet, a beautiful creation in latest chest-on-chest design. Carefully and ruggedly made of selected Sliced Walnut veneers with inlaid Sapeli veneer. Reverse curved front contour of selected hardwoods. Hand rubbed and finished to add lasting beauty. Size, 39 inches high, 28½ in. wide, and 12 in. deep.

Listed by Underwriters Laboratories and approved by the Radio Manufacturers' Association. For 100 to 125-volt 50 to 60-cycle A.C. If you live in a 25-cycle area add $4 and specify on order. Not Mailable; specify freight or express.

57 KM 7048—8-Tube Electric Radio. Shpg. wt., 72 lbs........................$49.95
Easy Payment Price ($5 down, $5 a month)............................ 54.35

DE LUXE CONCERT GRAND

10 Tubes Including Power Rectifier Tube
Truly a magnificent instrument! Brings you the unforgettable thrill and lasting pride of owning one of today's finest radios. Workmanship, sensitivity, tonal quality and fidelity approach perfection far beyond anything you have ever dreamed! A beautiful creation of the most skilled design and careful craftsmanship in a distinctively modern cabinet of refined, graceful simplicity. Elsewhere you would pay $99.50 for similar quality, but at Sears the price is based on honest dollar value with no additions to cover "trade-ins" or high salesmen's commissions. All the power and performance of this finest of all Silvertone Electric Consoles is yours for just $6 per month on our Easy Payment Plan.

$59.95 Cash
$5 DOWN

A Powerful 10-Tube Radio with all the sensitivity and selectivity that only a superheterodyne circuit and a 3-gang condenser can give you. As you tune distant stations the new Low Noise Converter practically eliminates the hissing and background noises. Has **4 Tuning Bands** of which 2 are our famous S-p-r-e-a-d Bands for easy reception of foreign stations at 9.4 to 9.8 MC. and 11.0 to 12.0 MC. Domestic and foreign shortwave stations are tuned from 6 to 18 MC. Regular American stations broadcast on the 540 to 1700 KC. range.

Big 12-Inch Dynamic Speaker that preserves the full richness of every instrument and every note. All the beauty of the world's finest music is yours to control because the Variable Tone Knob and the Bass Boost and Treble Boost Push Buttons work together to give you an infinite variety of tonal ranges—from the deep richness of the bass viol to the silver-pure intensity of a world-famous soprano. The tone is always full and mellow because of Automatic Bass Compensation. Volume is kept constant by the Automatic Volume Control.

More Beauty, More Power, More Features

Two Built-in Aerial Loops of newest low impedance type are also **Push Button Controlled!** One aerial is set to receive with maximum strength the broadcasting stations to the East or West; while the other aerial is set for station signals from North or South. Just touch the button to get clearer and stronger reception. However, where built-in aerial reception is not satisfactory, and for short-wave reception, an outside aerial must be attached to connections provided. (See Page 845 for Outside Aerial Kits.)

Eight Push Buttons in All. Two are Tone Booster Buttons and one is the Aerial Switching Control Button as described above. The other five buttons are for accurate easy tuning of any five favorite stations. Just push a button and there's your station—perfectly tuned. Easy to set-up too—it takes just 5 minutes with a screwdriver. Plug-in connection at back of radio provides for easy installation of phonograph record player, television or frequency modulation converter.

Gorgeous Console Cabinet of sliced walnut veneers and Sapeli wood inlays . . . a masterpiece of craftsmanship and skilled design. The recessed and slightly tilted instrument panel is set off by curved tambour-effect pilasters and the harmonious speaker grille bars are arranged in organ effect. Superbly built and finished in the expensive hand-rubbed manner usually found only in finest furniture. Size, 40 in. high, 29 in. wide, and 13¼ in. deep. Listed by Underwriters Laboratories. For 100 to 125-volt, 50 to 60-cycle A.C. If you live in a 25-cycle area add $5 and specify on order. Not Mailable; shipped freight or express.

57 KM 7050—10-Tube Electric Radio. Shpg. wt., 81 lbs. Cash............$59.95
Easy Payment Price ($5 down, $6 a month)........................... 65.45
For Easy Terms, use Easy Payment Side of Order Blank in Back of Book.

Items on these pages shipped from Mail Order Houses in Chicago, Philadelphia and Boston. 25-cycle models shipped from Chicago and Philadelphia only. Send order to nearest Mail Order House.

⊖ PAGE 837 . . . . RADIOS

Sonora *Clear as a Bell* A NEW HIGH IN RADIO VALUE

Sonora-Model TZ-56—12 TUBE AC CONSOLE

Nowhere is Sonora's rare ability to produce fine radio instruments at unbelievably low prices shown to better advantage than in this remarkable 12-tube console. From "stem to stern" the Model TZ-56 fairly glistens with quality and precision. 3 full wave bands—6 station automatic tuning—Built-in Sonorascope—these are but a few of the many outstanding features offered in this model. We urge you to compare this set, feature for feature, with other competitive brands of radios, selling in the same price range. Only then will you have a true picture of the remarkable value which Sonora is placing before you. You can sell this set to your customers with the understanding that, if after having used it for 30 days, they are not entirely convinced that it is an outstanding value, they can return it and receive full refund. We will back you in this offer 100%.

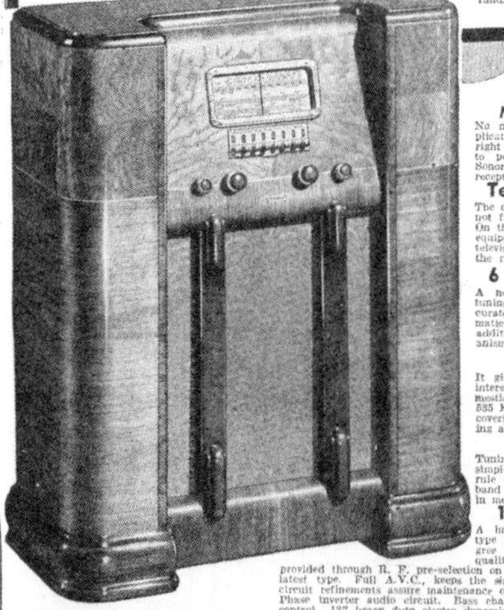

Built-In "Sonorascope"
No Aerial or Ground Required

No messy wires stretching across floors—no complicated installations. The "Sonorascope" built right into the set itself eliminates all that. Brought to perfection by Sonora's engineering staff, the Sonorascope provides the utmost in far-reaching reception on all wave bands.

Television-Phono Connection

The coming of television on a widespread scale will not find the Sonora TZ-56 out-of-date or obsolete. On the contrary, this receiver is ready for it, being equipped with a television connection into which a television unit can be plugged for operation with the radio. A phonograph jack is also provided.

6 Station Automatic Tuning

A newly perfected, electrically driven automatic tuning mechanism, easy to adjust and 100% accurate. Any 6 stations can be selected for automatic tuning by the mere touch of a button. An additional button is for converting the tuning mechanism back to manual tuning.

3 Full Wave Bands

It gives the listener a lifetime pass to anything interesting on the airways, foreign as well as domestic. The wave bands are as follows: 1720 to 535 K.C. for standard broadcast; 1710 to 5650 K.C. covering police short wave; 5.65 to 18.1 M.C. covering all foreign channels.

8½" Slide Rule Dial

Tuning on all wave bands is made exceedingly simple by means of the 8½x4" full-vision slide rule dial with two-tone gold leaf face. Broadcast band is calibrated in kilocycles; short wave band in megacycles.

12-In. Auditorium Speaker

A brand new superheterodyne in which 12 new type tubes are used to provide a remarkable degree of sensitivity and "Clear as a Bell" tone quality. More dependable distance tuning is provided through R. F. pre-selection on broadcast band. Three gang condenser of the latest type. Full A.V.C., keeps the signal intensity equal on all stations. Special circuit refinements assure maintenance of permanent stability in all oscillator circuits. Phase inverter audio circuit. Bass channel amplifier and continuously variable tone control. 12" heavy duty electro dynamic speaker. 7½ watt output. Uses the following tubes: 1—6A8G, 7—6J5G, 1—6K7G, 2—6F6G and 1—80.

An Engineering Masterpiece

Sonora engineers spent many hours of exhaustive research before finally hitting upon the design of this 12-tube superheterodyne receiver. However, the results have more than justified the efforts. It is safe to say that here indeed is radio at its finest.

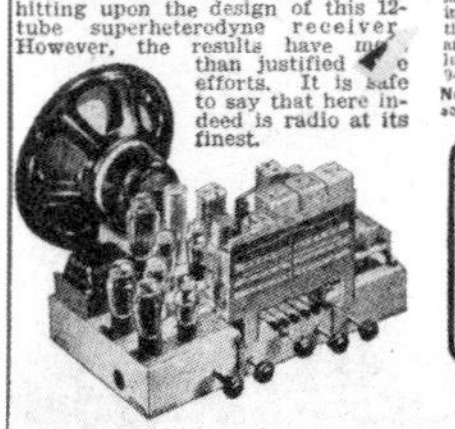

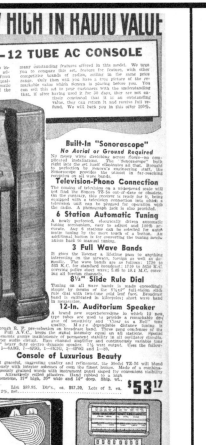

Console of Luxurious Beauty

Massive, yet graceful, suggesting quality and refinement, the Model TZ-56 will blend in harmoniously with interior schemes of even the finest homes. Made of a combination of gorgeously grained woods with instrument panel sloped for convenient visibility and tuning. Attractively rolled pilasters. Hand rubbed to a high lustre. Dimensions, 37" high, 30" wide and 14" deep. Ship. wt., 94½ lbs.

No. 9920. List $89.95. Dlr's., ea. $57.20. Lots of 3, ea. $54.25, less 2%, net.......... **$53.17**

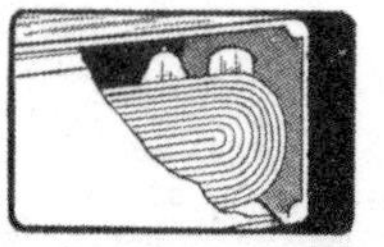

Built-in Sonorascope—No Aerial or Ground Required.

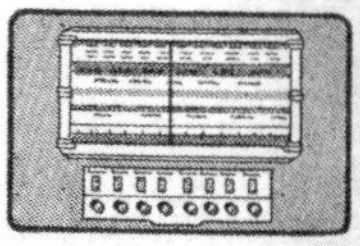

8½"x4" Full Vision Slide Rule Dial.

1940 Catalog Page

Sonora *Clear as a Bell* *Perfected* PUSH BUTTON TUNING

Sonora-Model TT-52
5 TUBE AC SUPERHET
8 Tube Performance!

Here is a cleverly designed Sonora model that fairly radiates quality and good taste. Those of your customers seeking a powerful, yet compact receiver with which they can enjoy thoroughly dependable radio performance and decorative beauty will certainly voice their preference for the Model TT-52.

As low as **$15.40**

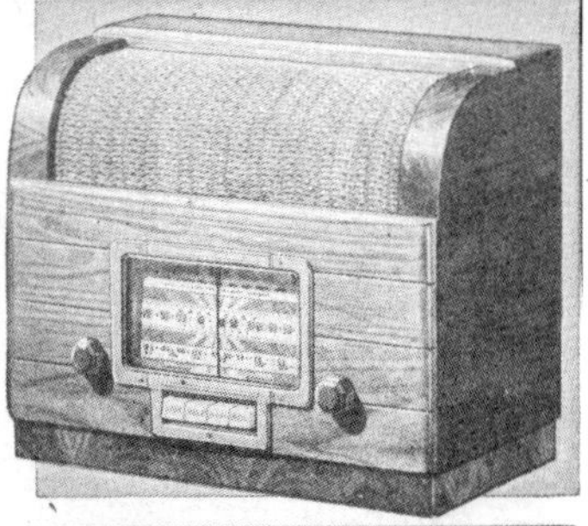

Perfected Automatic Tuning

Any 4 stations can be selected for automatic push button selection. Perfected automatic tuning mechanism makes tuning entirely foolproof and effortless. Only ⅜" stroke required. Magni-Glas button face provides increased visibility of station call letters.

Television Phono Connections

Precision engineered throughout by Sonora's expert engineering staff, the superheterodyne circuit employed in the model TT-52 leaves nothing to be desired in the way of tone quality, selectivity and undistorted volume. Dual purpose and triple purpose tubes are used, thus giving the performance quality of 8 and 9 tube receivers. Automatic volume control. Has a remarkable output of 2½ watts. Tubes are as follows: 1—6A8GT, 1—6K7GT, 1—6Q7GT, 1—6K6 and 1—80. Equipped with **television and phonograph connections.**

No Aerial or Ground Required

This model features the famous built-in "Sonorascope" and as a result, the necessity for aerial or ground connections is eliminated. Just plug into electrical outlet and tune. Tunes all standard broadcast programs on 1720 to 535 K.C., including 1712 K.C. Police Channel. 5¾" by 4⅞" Gem-loid tri-colored slide rule dial. New type plastic integral one piece dial face and escutcheon. Equipped with 5" electro dynamic speaker.

Handsome New Upright Cabinet

This strikingly modern walnut cabinet is fashioned of luxuriously grained walnut veneers and features a unique and attractive waterfall speaker grille design. Dimensions, 12½" wide, 10½" high, 7⅝" deep. Operates on 110 volts AC only. Ship. wt., 12½ lbs.

No. 9916. List $24.95. Dlr's., ea. $16.63. Lots of 3, ea. $15.71, less 2%, net.......... **15.40**

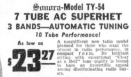

JUST PRESS A BUTTON!

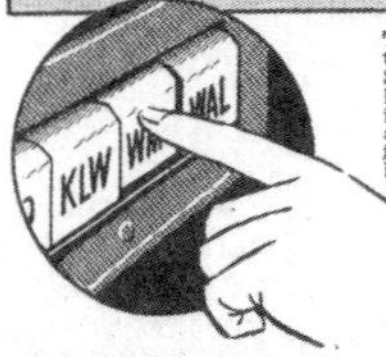

The improved automatic tuning mechanism on Sonora radios is amazingly simple both as to initial adjustment and operation. The station desired for automatic selection can be quickly adjusted by anyone—right from the front. No complicated hookup — only a screw driver required. Touch-button selection is effortless — a light ⅜" stroke does the trick.

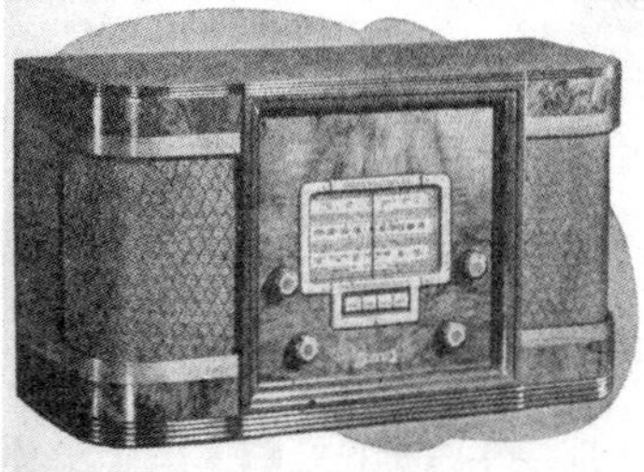

Built-In "Sonorascope"—No Aerial or Ground Required

The efforts of Sonora's entire engineering staff has gone into the planning of the superheterodyne circuit and the results are at once apparent in the superb tone quality and extreme sensitivity which characterizes this model. Equipped with built-in "Sonorascope"—no aerial or ground required. Automatic volume control. Continuously variable tone control. 6" electro dynamic speaker. 3½ watt output. **Television and phonograph connections.** Tubes are as follows: 1-6A8G, 1-6K7G, 1-6Q7G, 1-6J5G, 1-6K6G and 1-80.

No. 9018. List $43.95. Dlr's., ea. $24.85. Lots of 3, ea. $23.75, less 2%, net.......... **23.27**

Sonora-Model TY-54
7 TUBE AC SUPERHET
3 BANDS—AUTOMATIC TUNING
10 Tube Performance!

As low as **$23.27**

A magnificent new table model planned for those who want the utmost in radio performance, in compact form. Its brilliant styling with exceptional "Clear as a Bell" tone quality is bound to have an irresistible appeal among discriminating radio buyers.

Foreign Reception

Three wave bands provide the listener with everything worth while in radio reception. 1720 to 535 K. C.'s cover the standard broadcast band; 7.5 to 2.2 MC cover police channels; 24 to 7.25 MC covers all Foreign Channels including the popular new 13 meter band. All three bands are clearly calibrated on 5¾ by 4⅞ Gem-Loid tri-colored slide rule dial.

4 Button Automatic Tuning

Any 4 stations can be easily and quickly adjusted for automatic push-button selection. Newly developed Sonora automatic tuning mechanism features effortless ⅜" stroke and Magni-Glass button face, giving increased visibility of station call letters.

Unique Cabinet Design

A glorious combination of luxuriously grained veneers form a cabinet of striking beauty. The clever arrangement of the dual speaker grille enhances the individuality of this cabinet. Hand rubbed to a high lustre finish. Dimensions, 20½" wide, 11⅞" high, 8¼" deep. Ship. wt. 19¼ lbs. Operates on 110 volt, AC only.

1940 Catalog Page

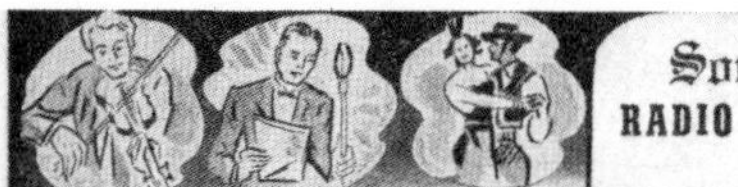

Sonora MODEL KXF-95
RADIO PHONO COMBINATION

- **AUTOMATIC RECORD CHANGER**
  Plays eight 10" or seven 12" records at a single loading
- **FOREIGN RECEPTION**
  2 full wave bands—gets interesting foreign as well as domestic broadcasts
- **AUTOMATIC PUSH-BUTTON TUNING**
  Any 6 stations can be tuned automatically—at the touch of a button

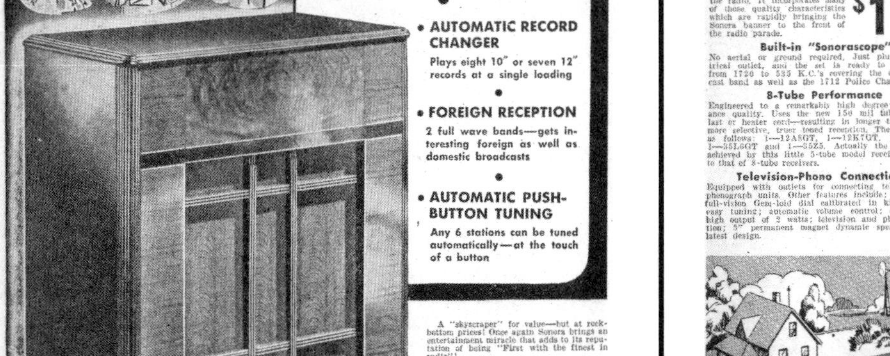

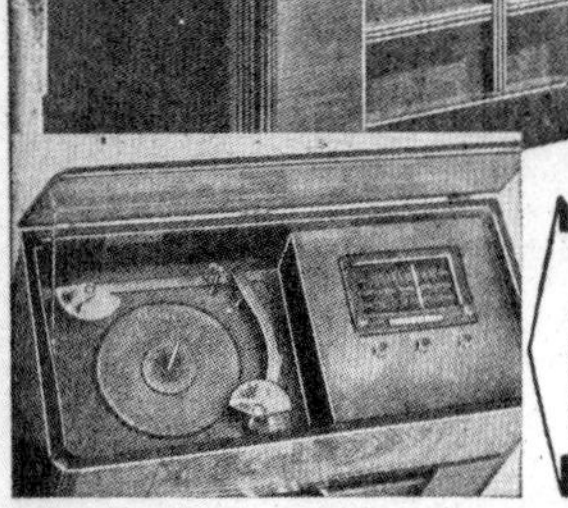

The Inside Story

Sonora engineers designed this model as a fine instrument for human enjoyment as well as a piece of furniture that will add a note of luxury to any home. From inside or outside, it spells up-to-the-second achievements in performance and looks!

A "skyscraper" for value—but at rock-bottom prices! Once again Sonora brings an entertainment miracle that adds to its reputation of being "First with the finest in radio"!

Breath-taking New Beauty—with the fine simplicity of line that is a masterful combination of ancient and modern styles of design. This rich-looking console model proclaims the good taste and smartness of its owner. Be sure to feature it in your radio displays. When you offer a model such as this to your customers — they instinctively feel that you flatter their sense of appreciation for the finer things in life. For, this model is one of the finer things that can grace their homes. It is made of choice, skillfully matched, Walnut veneers with artistic piano finish.

Made For Critical radio listeners and music lovers who demand the best. This Console Grande Combination spells "front row" enjoyment for every single performance — whether musical, phonograph, dramatic or news broadcast. This is truly a luxury instrument. Be sure to "talk up" this "luxury" angle in selling prospective customers!

See Exciting Features below. Have you ever heard of a better buy in Radio-Phono combinations than this? It's truly a Sonora triumph!

Built-In "Sonorascope" — No Aerial or Ground Needed!

Famous enclosed aerial loop makes this fine console combination just another electrical appliance around the house. There's no messy wires to stretch across the floor—no complicated installations to set the tempers boiling. Just plug it in—and hear it play! Extra connection for outside aerial also included.

Rare Perfection of Tone Means Rare Enjoyment!

Only the richness and naturalness of the radio program as it leaves the studio comes over this Console Grande Combination. Special tone control means a more sensitive radio performance keyed to bring greater pleasure to listeners. Automatic volume control keeps signal intensity equal on all stations.

Latest Superhet Design — 9 Tube Performance

A brand new Superheterodyne with 6 new 1941 tubes brings the well-known "Clear As a Bell" tone that your customers naturally expect from a Sonora radio. Five watt output. Tubes are as follows: 1—6SA7GT, 1—6SK7BT, 1—6SQ7GT, 1—25C6G, 2—25Z6GT.

Rich-Sounding Phono Tones, Too!

Phonograph has self-starting Squirrel Cage induction-type motor that operates silently and maintains constant speed. Plays records with lid up or down. Automatic Record Changer is a real luxury — you can listen for thirty minutes without touching the instrument. Crystal pickup. Plays eight 10" or seven 12" records at a single loading! Tell your customers they can select their records—then relax.

Foreign or American Radio Reception

Thrilling two-band reception brings favorite radio programs from home or abroad. This is a big point in selling radios today! Americans everywhere are intensely interested in happenings in the outside world. This luxurious radio lets them hear on-the-spot news broadcasts from the capitals of Europe, South America, and Africa as well as enjoyable domestic, musical and dramatic programs. Tunes 535-1720 K.C. (for domestic stations and popular 1712 Police Channel) and 5.65 - 18.3 M.C. (for short-wave foreign stations). 110 volts AC, 60 cycles.

Every Imaginable Convenience Feature Included!

To really enjoy oneself, one must be in a state of complete relaxation. Sonora engineers had this in mind when they designed this model. Large 9" x 6" Slide-Rule dial makes hand-dialing easier. Special Push-Button Tuning permits a finger's touch to bring any one of six favorite stations. Pilot-light "Off-On" Indicator. **Size of model:** 33¾" Long, 32" High, 17" Deep. Shipping weight 105 lbs.

No. 9977. List price $99.95. Dlr's., ea. $55.49, lots of 3, ea. $52.50, less 2%, net.......... **51.45**

1941 Catalog Page

Sonora-Model TJ-63
5 TUBE AC-DC SUPERHET

For those preferring a small, simple, wooden cabinet receiver, you will find the TJ-63 just the radio. It incorporates many of those quality characteristics which are rapidly bringing the Sonora banner to the front of the radio parade.

As Low As **$12.88**

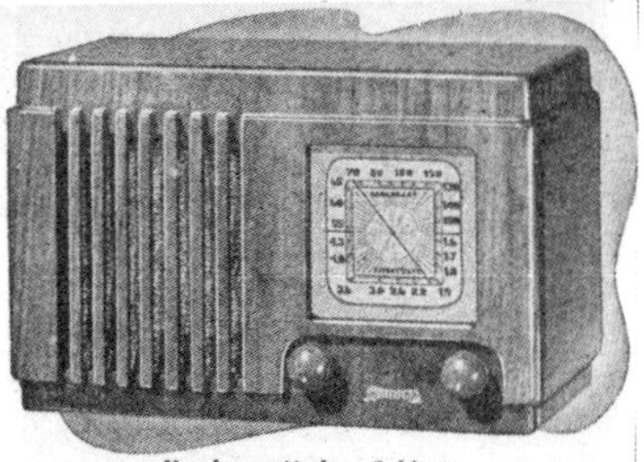

Built-in "Sonorascope"

No aerial or ground required. Just plug into electrical outlet, and the set is ready to tune. Tunes from 1720 to 535 K.C.'s covering the entire broadcast band as well as the 1712 Police Channel

8-Tube Performance

Engineered to a remarkably high degree of performance quality. Uses the new 150 mil tubes—no ballast or heater cord—resulting in longer tube life and more selective, truer toned reception. These tubes are as follows: 1—12A8GT, 1—12K7GT, 1—12Q7GT, 1—35L6GT and 1—35Z5. Actually the performance achieved by this little 5-tube model receiver is equal to that of 8-tube receivers.

Television-Phono Connections

Equipped with outlets for connecting television and phonograph units. Other features include: New square full-vision Gem-loid dial calibrated in kilocycles for easy tuning; automatic volume control; exceptionally high output of 2 watts; television and phono connection; 5" permanent magnet dynamic speaker of the latest design.

Handsome Modern Cabinet

Pleasingly modern and in exceedingly good taste is this handsome little table cabinet of solid walnut in harmonious two-tone effect. Unique type of speaker grille opening. The rich effect of the Gem-loid dial blends in beautifully with the entire cabinet. Dimensions of cabinet, 12" wide, 7¼" high, 7" deep. Ship. wt., 9¼ lbs. Operates on 110-120 volts, 40-60 cycles, AC or DC.

No. 9914. List $20.95. Dlr's., ea. $13.67. Lots of 3, ea. $13.14, less 2%, net.......... **12.88**

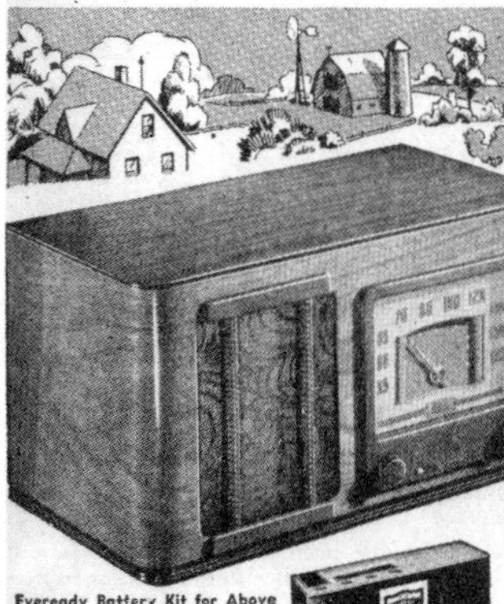

Economy on the Farm!
Sonora MODEL KZ-111
1.4 VOLT BATTERY RECEIVER
4 Tube Superhet

- Low Battery Consumption
- 7 Tube Performance
- Clock-Type Glass Dial
- Automatic Volume Control
- Walnut Veneer Cabinet
- 6" P.M. Dynamic Speaker
- "On-Off" Visual Switch

When economy means real entertainment too - - - your customers out on the farm are sure to sit up and take notice! We know of no other Farm Radio on the market today that combines such downright thriftiness and far-reaching power as the handsome SONORA model!

Just Like a "Big City" Radio! SONORA engineers have taken great pains to give this model the smooth performance and style appeal of "city" radios. Your eye will spot the difference in a second - - - rich-looking Walnut veneer cabinet of table size! Makes friends at a glance—and keeps them through perfect performance. Large 7" easy-reading glass dial. Convenient "On-Off" Visual switch. Tuning range from 535-1720 K.C. (entire broadcast band and 1712 Police Channel). Powerful 6" Dynamic Speaker. 2 gang condenser of latest improved type. Full automatic volume control keeps the signal intensity equal on all stations.

All the thrills of 7 tube performance at lower than 4-Tube prices! Latest style Superheterodyne circuit. 250 milliwatt output. Contains the following tubes: 1 — 1A7GT, 1 — 1N5GT, 1—1H5GT, 1—1Q5GT.

Beautifully proportioned table model cabinet of selected, richly grained walnut is in keeping with the beauty which pours forth from its speaker grille. Ample space in rear for battery pack. Size 17" wide, 10½" high, 9¾" deep. Ship. wt. 12¾ lbs.

No. 9985. List $20.95, complete with tubes, less battery. Dlr's., ea. $13.06, lots of 3, ea. $12.59, less 2%, net.......... **12.34**

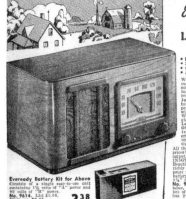

Eveready Battery Kit for Above

Consists of a single easy-to-use unit containing 1½ volts of "A" power and 90 volts of "B" power.

No. 9614. List $5.00, Dlr's., ea. $3.45, less 2%, net.......... **3.38**

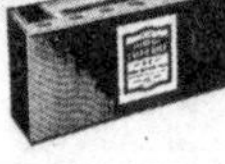

GO AFTER RADIO PROFITS THE MODERN WAY . . . THE Sonora WAY

Dealers who have taken on the Sonora line have received a pleasant surprise at the success they have had with it. But to us, who have watched Sonora grow from very humble beginnings to its present dominant position in the radio field, it is really no surprise at all. We knew that Sonora was going places when we saw how well Sonora was meeting the rapidly changing radio trends. That's the whole idea behind Sonora's success—they keep a constant finger on the pulse of public demand. Every model has a definite place in the Sonora scheme. There simply are no "duds". Let us tell you more about this fast-moving radio line. Write in and tell us "you're Sonora-conscious".

1941 Catalog Page

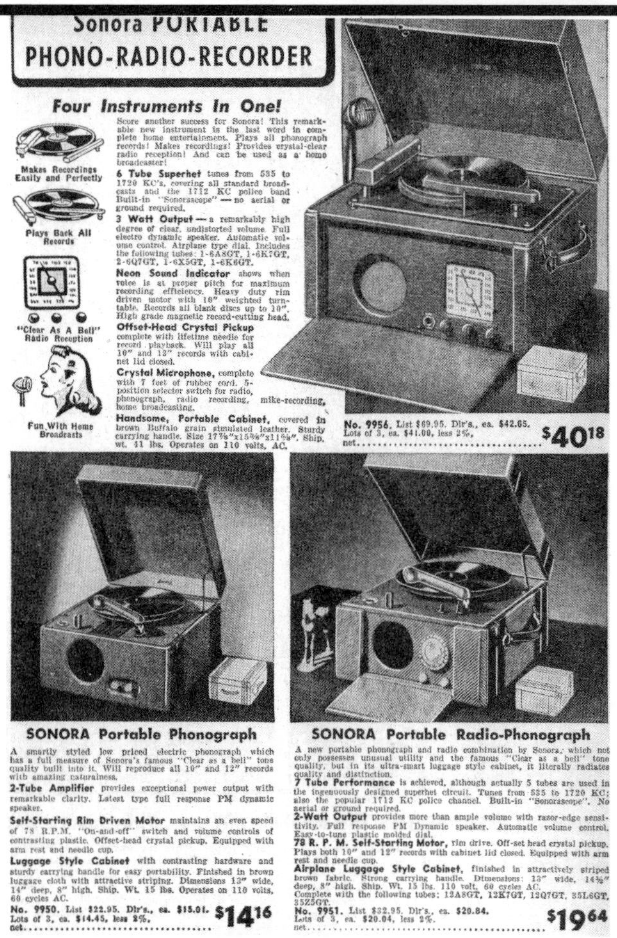

Sonora PORTABLE PHONO-RADIO-RECORDER

Four Instruments In One!

Score another success for Sonora! This remarkable new instrument is the last word in complete home entertainment. Plays all phonograph records! Makes recordings! Provides crystal-clear radio reception! And can be used as a home broadcaster!

6 Tube Superhet tunes from 535 to 1720 KC's, covering all standard broadcasts and the 1712 KC police band Built-in "Sonorascope" — no aerial or ground required.

3 Watt Output — a remarkably high degree of clear, undistorted volume. Full electro dynamic speaker. Automatic volume control. Airplane type dial. Includes the following tubes: 1-6A8GT, 1-6K7GT, 2-6Q7GT, 1-6X5GT, 1-6K6GT.

Neon Sound Indicator shows when voice is at proper pitch for maximum recording efficiency. Heavy duty rim driven motor with 10" weighted turntable. Records all blank discs up to 10". High grade magnetic record-cutting head.

Offset-Head Crystal Pickup complete with lifetime needle for record playback. Will play all 10" and 12" records with cabinet lid closed.

Crystal Microphone, complete with 7 feet of rubber cord. 5-position selector switch for radio, phonograph, radio recording, mike-recording, home broadcasting.

Handsome, Portable Cabinet, covered in brown Buffalo grain simulated leather. Sturdy carrying handle. Size 17⅞"x15⅝"x11⅝". Ship. wt. 41 lbs. Operates on 110 volts, AC.

Makes Recordings Easily and Perfectly

Plays Back All Records

"Clear As A Bell" Radio Reception

Fun With Home Broadcasts

No. 9956. List $69.95. Dlr's., ea. $42.65. Lots of 3, ea. $41.00, less 2%. net........ $40.18

SONORA Portable Phonograph

A smartly styled low priced electric phonograph which has a full measure of Sonora's famous "Clear as a bell" tone quality built into it. Will reproduce all 10" and 12" records with amazing naturalness.

2-Tube Amplifier provides exceptional power output with remarkable clarity. Latest type full response PM dynamic speaker.

Self-Starting Rim Driven Motor maintains an even speed of 78 R.P.M. "On-and-off" switch and volume controls of contrasting plastic. Offset-head crystal pickup. Equipped with arm rest and needle cup.

Luggage Style Cabinet with contrasting hardware and sturdy carrying handle for easy portability. Finished in brown luggage cloth with attractive striping. Dimensions 13" wide, 14" deep, 8" high. Ship. Wt. 15 lbs. Operates on 110 volts, 60 cycles AC.

No. 9950. List $22.95. Dlr's., ea. $15.01. Lots of 3, ea. $14.45, less 2%. net........ $14.16

SONORA Portable Radio-Phonograph

A new portable phonograph and radio combination by Sonora, which not only possesses unusual utility and the famous "Clear as a bell" tone quality, but in its ultra-smart luggage style cabinet, it literally radiates quality and distinction.

7 Tube Performance is achieved, although actually 5 tubes are used in the ingenuously designed superhet circuit. Tunes from 535 to 1720 KC; also the popular 1712 KC police channel. Built-in "Sonorascope". No aerial or ground required.

2-Watt Output provides more than ample volume with razor-edge sensitivity. Full response PM Dynamic speaker. Automatic volume control. Easy-to-tune plastic molded dial.

78 R. P. M. Self-Starting Motor, rim drive. Off-set head crystal pickup. Plays both 10" and 12" records with cabinet lid closed. Equipped with arm rest and needle cup.

Airplane Luggage Style Cabinet, finished in attractively striped brown fabric. Strong carrying handle. Dimensions: 13" wide, 14⅜" deep, 8" high. Ship. Wt. 15 lbs. 110 volt, 60 cycles AC. Complete with the following tubes: 12A8GT, 12K7GT, 12Q7GT, 35L6GT, 35Z5GT.

No. 9951. List $32.95. Dlr's., ea. $20.84. Lots of 3, ea. $20.04, less 2%. net........ $19.64

1941 Catalog Page

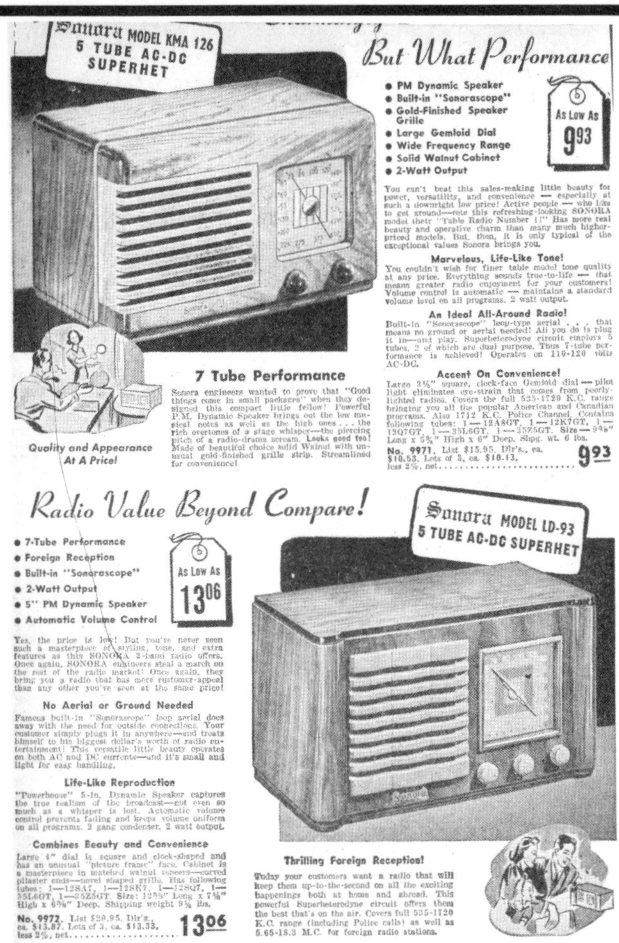

Sonora MODEL KMA 126 5 TUBE AC-DC SUPERHET

But What Performance

- PM Dynamic Speaker
- Built-in "Sonorascope"
- Gold-Finished Speaker Grille
- Large Gemloid Dial
- Wide Frequency Range
- Solid Walnut Cabinet
- 2-Watt Output

As Low As 9.93

You can't beat this sales-making little beauty for power, versatility, and convenience — especially at such a downright low price! Active people — who like to get around—vote this refreshing-looking SONORA model their "Table Radio Number 1!" Has more real beauty and operative charm than many much higher-priced models. But, then, it is only typical of the exceptional values Sonora brings you.

Marvelous, Life-Like Tone!
You couldn't wish for finer table model tone quality at any price. Everything sounds true-to-life — that means greater radio enjoyment for your customers! Volume control is automatic — maintains a standard volume level on all programs. 2 watt output.

An Ideal All-Around Radio!
Built-in "Sonorascope" loop-type aerial . . . that means no ground or aerial needed! All you do is plug it in—and play. Superheterodyne circuit employs 5 tubes, 2 of which are dual purpose. Thus 7-tube performance is achieved! Operates on 110-120 volts AC-DC.

Accent On Convenience!
Large 3½" square, clock-face Gemloid dial — pilot light eliminates eye-strain that comes from poorly-lighted radios. Covers the full 535-1720 K.C. range bringing you all the popular American and Canadian programs. Also 1712 K.C. Police Channel. Contains following tubes: 1—12A8GT, 1—12K7GT, 1—12Q7GT, 1—35L6GT, 1—35Z5GT. Size—9⅜" Long x 5⅝" High x 6" Deep. Shpg. wt. 6 lbs.

No. 9971. List $15.95. Dlr's., ea. $10.53. Lots of 3, ea. $10.13, less 2%, net........ 9.93

Quality and Appearance At A Price!

7 Tube Performance
Sonora engineers wanted to prove that "Good things come in small packages" when they designed this compact little fellow! Powerful P.M. Dynamic Speaker brings out the low musical notes as well as the high ones . . . the rich overtones of a stage whisper—the piercing pitch of a radio-drama scream. Looks good too! Made of beautiful choice solid Walnut with unusual gold-finished grille strip. Streamlined for convenience!

Radio Value Beyond Compare!

Sonora MODEL LD-93 5 TUBE AC-DC SUPERHET

- 7-Tube Performance
- Foreign Reception
- Built-in "Sonorascope"
- 2-Watt Output
- 5" PM Dynamic Speaker
- Automatic Volume Control

As Low As 13.06

Yes, the price is low! But you've never seen such a masterpiece of styling, tone, and extra features as this SONORA 2-band radio offers. Once again, SONORA engineers steal a march on the rest of the radio market! Once again, they bring you a radio that has more customer-appeal than any other you've seen at the same price!

No Aerial or Ground Needed
Famous built-in "Sonorascope" loop aerial does away with the need for outside connections. Your customer simply plugs it in anywhere—and treats himself to his biggest dollar's worth of radio entertainment! This versatile little beauty operates on both AC and DC currents—and it's small and light for easy handling.

Life-Like Reproduction
"Powerhouse" 5-In. Dynamic Speaker captures the true realism of the broadcast—not even so much as a whisper is lost. Automatic volume control prevents fading and keeps volume uniform on all programs. 2 gang condenser, 2 watt output.

Combines Beauty and Convenience
Large 4" dial is square and clock-shaped and has an unusual "picture frame" face. Cabinet is a masterpiece in matched walnut veneers—curved pilaster ends—novel shaped grille. Has following tubes: 1—12SA7, 1—12SK7, 1—12SQ7, 1—35L6GT, 1—35Z5GT. Size: 12⅝" Long x 7⅛" High x 6⅝" Deep. Shipping weight 9½ lbs.

No. 9972. List $20.95. Dlr's., ea. $13.87. Lots of 3, ea. $13.33, less 2%, net........ 13.06

Thrilling Foreign Reception!
Today your customers want a radio that will keep them up-to-the-second on all the exciting happenings both at home and abroad. This powerful Superheterodyne circuit offers them the best that's on the air. Covers full 535-1720 K.C. range (including Police calls) as well as 5.65-18.3 M.C. for foreign radio stations.

1941 Catalog Page

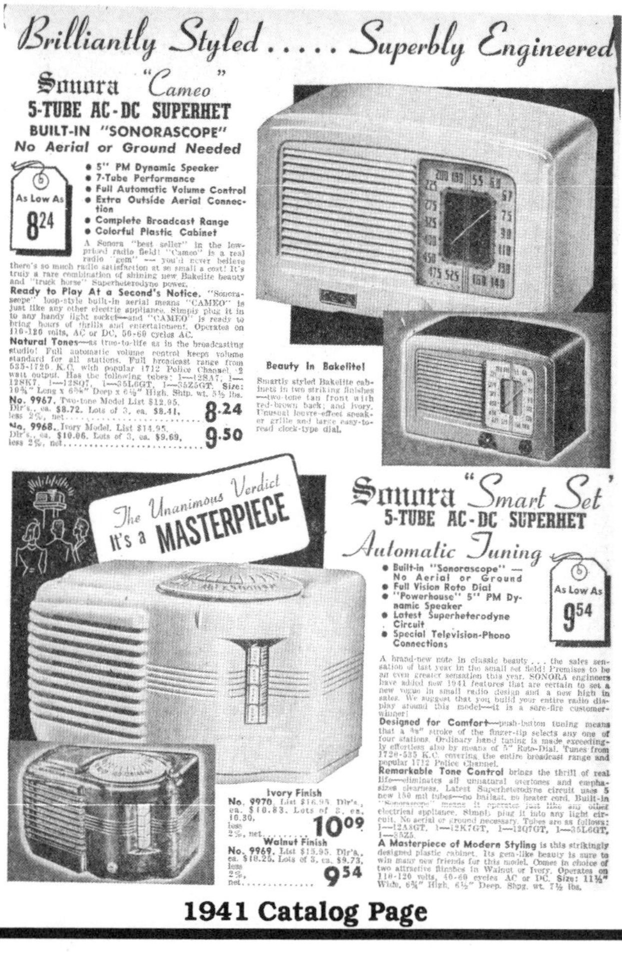

Brilliantly Styled . . . . . Superbly Engineered

Sonora "Cameo" 5-TUBE AC-DC SUPERHET
BUILT-IN "SONORASCOPE"
No Aerial or Ground Needed

As Low As 8.24

- 5" PM Dynamic Speaker
- 7-Tube Performance
- Full Automatic Volume Control
- Extra Outside Aerial Connection
- Complete Broadcast Range
- Colorful Plastic Cabinet

A Sonora "best seller" in the low-priced radio field! "Cameo" is a real radio "gem" — you'd never believe there's so much radio satisfaction at so small a cost! It's truly a rare combination of shining new Bakelite beauty and "truck horse" Superheterodyne power.

Ready to Play At a Second's Notice. "Sonorascope" loop-style built-in aerial means "CAMEO" is just like any other electric appliance. Simply plug it in to any handy light socket—and "CAMEO" is ready to bring hours of thrills and entertainment. Operates on 110-120 volts, AC or DC, 50-60 cycles AC.

Natural Tones—as true-to-life as in the broadcasting studio! Full automatic volume control keeps volume standard for all stations. Full broadcast range from 535-1720 K.C. with popular 1712 Police Channel. 2 watt output. Has the following tubes: 1—12SA7, 1—12SK7, 1—12SQ7, 1—35L6GT, 1—35Z5GT. Size: 10¾" Long x 6⅝" Deep x 6½" High. Ship. wt. 5½ lbs.

No. 9967. Two-tone Model List $12.95. Dlr's., ea. $8.72. Lots of 3, ea. $8.41, less 2%, net........ 8.24

No. 9968. Ivory Model. List $14.95. Dlr's., ea. $10.06. Lots of 3, ea. $9.69, less 2%, net........ 9.50

Beauty In Bakelite!
Smartly styled Bakelite cabinets in two striking finishes—two-tone tan front with red-brown back; and ivory. Unusual louvre-effect speaker grille and large easy-to-read clock-type dial.

The Unanimous Verdict It's a MASTERPIECE

Sonora "Smart Set" 5-TUBE AC-DC SUPERHET
Automatic Tuning

- Built-in "Sonorascope" — No Aerial or Ground
- Full Vision Roto Dial
- "Powerhouse" 5" PM Dynamic Speaker
- Latest Superheterodyne Circuit
- Special Television-Phono Connections

As Low As 9.54

A brand-new note in classic beauty . . . the sales sensation of last year in the small set field! Promises to be an even greater sensation this year. SONORA engineers have added new 1941 features that are certain to set a new vogue in small radio design and a new high in sales. We suggest that you build your entire radio display around this model—it is a sure-fire customer-winner!

Designed for Comfort—push-button tuning means that a ⅜" stroke of the finger-tip selects any one of four stations. Ordinary hand tuning is made exceedingly effortless also by means of 5" Roto-Dial. Tunes from 1720-535 K.C. covering the entire broadcast range and popular 1712 Police Channel.

Remarkable Tone Control brings the thrill of real life—eliminates all unnatural overtones and emphasizes clearness. Latest Superheterodyne circuit uses 5 new 150 mil tubes—no ballast, no heater cord. Built-in "Sonorascope" means it operates just like any other electrical appliance. Simply plug it into any light circuit. No aerial or ground necessary. Tubes are as follows: 1—12A8GT, 1—12K7GT, 1—12Q7GT, 1—35L6GT, 1—35Z5.

A Masterpiece of Modern Styling is this strikingly designed plastic cabinet. Its gem-like beauty is sure to win many new friends for this model. Comes in choice of two attractive finishes in Walnut or Ivory. Operates on 110-120 volts, 40-60 cycles AC or DC. Size: 11½" Wide, 6¾" High, 6½" Deep. Shpg. wt. 7½ lbs.

Ivory Finish
No. 9970. List $16.95. Dlr's., ea. $10.83. Lots of 3, ea. 10.30, less 2%, net........ 10.09

Walnut Finish
No. 9969. List $15.95. Dlr's., ea. $10.25. Lots of 3, ea. $9.73, less 2%, net........ 9.54

1941 Catalog Page

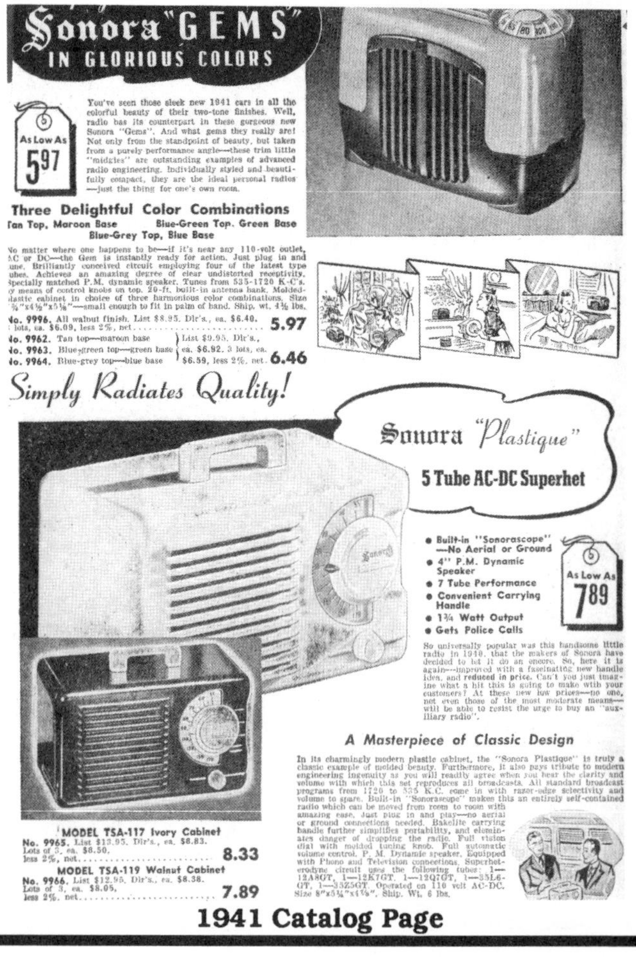

Sonora "GEMS" IN GLORIOUS COLORS

As Low As 5.97

You've seen those sleek new 1941 cars in all the colorful beauty of their two-tone finishes. Well, radio has its counterpart in these gorgeous new Sonora "Gems". And what gems they really are! Not only from the standpoint of beauty, but taken from a purely performance angle—these trim little "midgies" are outstanding examples of advanced radio engineering. Individually styled and beautifully compact, they are the ideal personal radios—just the thing for one's own room.

Three Delightful Color Combinations
Tan Top, Maroon Base — Blue-Green Top, Green Base — Blue-Grey Top, Blue Base

No matter where one happens to be—if it's near any 110-volt outlet, AC or DC—the Gem is instantly ready for action. Just plug in and tune. Brilliantly conceived circuit employing four of the latest type tubes. Achieves an amazing degree of clear undistorted receptivity. Specially matched P.M. dynamic speaker. Tunes from 535-1720 K-C's. by means of control knobs on top. 20-ft. built-in antenna bank. Molded-plastic cabinet in choice of three harmonious color combinations. Size ¾"x4½"x5⅛"—small enough to fit in palm of hand. Ship. wt. 4½ lbs.

No. 9996. All walnut finish. List $8.95. Dlr's., ea. $6.40. 3 lots, ea. $6.09, less 2%, net........ 5.97

No. 9962. Tan top—maroon base
No. 9963. Blue-green top—green base
No. 9964. Blue-grey top—blue base
List $9.95. Dlr's., ea. $6.92. 3 lots, ea. $6.59, less 2%, net. 6.46

Simply Radiates Quality!

Sonora "Plastique" 5 Tube AC-DC Superhet

- Built-in "Sonorascope" —No Aerial or Ground
- 4" P.M. Dynamic Speaker
- 7 Tube Performance
- Convenient Carrying Handle
- 1¾ Watt Output
- Gets Police Calls

As Low As 7.89

So universally popular was this handsome little radio in 1940, that the makers of Sonora have decided to let it do an encore. So, here it is again—improved with a fascinating new handle idea, and **reduced in price.** Can't you just imagine what a hit this is going to make with your customers? At these new low prices—no one, not even those of the most moderate means—will be able to resist the urge to buy an "auxiliary radio".

A Masterpiece of Classic Design
In its charmingly modern plastic cabinet, the "Sonora Plastique" is truly a classic example of molded beauty. Furthermore, it also pays tribute to modern engineering ingenuity as you will readily agree when you hear the clarity and volume with which this set reproduces all broadcasts. All standard broadcast programs from 1720 to 535 K.C. come in with razor-edge selectivity and volume to spare. Built-in "Sonorascope" makes this an entirely self-contained radio which can be moved from room to room with amazing ease. Just plug in and play—no aerial or ground connections needed. Bakelite carrying handle further simplifies portability, and eliminates danger of dropping the radio. Full vision dial with molded tuning knob. Full automatic volume control. P. M. Dynamic speaker. Equipped with Phono and Television connections. Superheterodyne circuit uses the following tubes: 1—12A8GT, 1—12K7GT, 1—12Q7GT, 1—35L6GT, 1—35Z5GT. Operated on 110 volt AC-DC. Size 8"x5¾"x4⅞". Ship. Wt. 6 lbs.

MODEL TSA-117 Ivory Cabinet
No. 9965. List $13.95. Dlr's., ea. $8.83. Lots of 3, ea. $8.50, less 2%, net........ 8.33

MODEL TSA-119 Walnut Cabinet
No. 9966. List $12.95. Dlr's., ea. $8.38. Lots of 3, ea. $8.05, less 2%, net........ 7.89

1941 Catalog Page

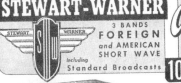

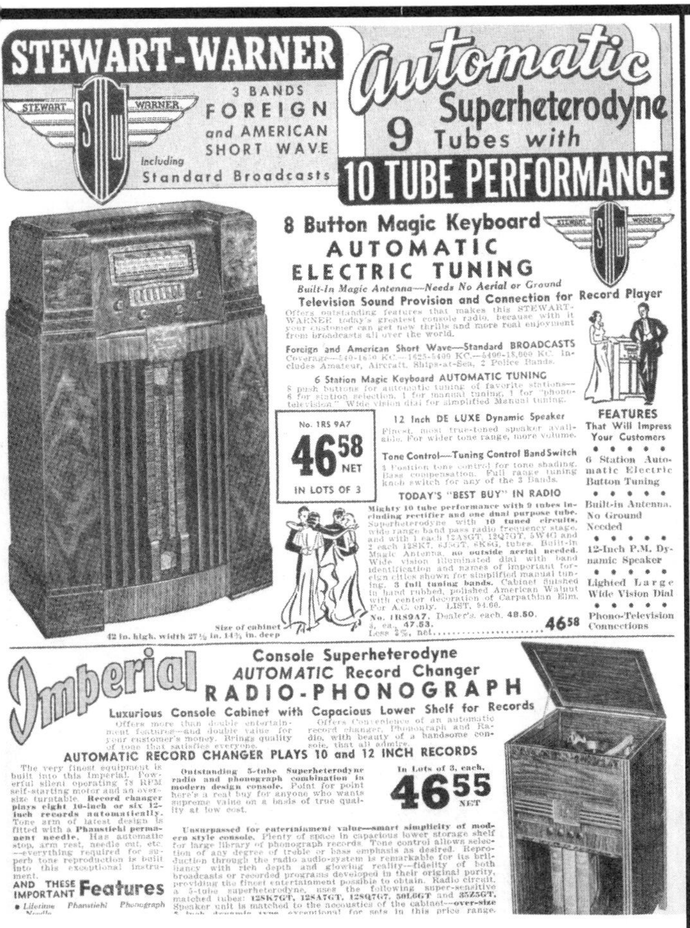

STEWART-WARNER

3 BANDS FOREIGN and AMERICAN SHORT WAVE Including Standard Broadcasts

Automatic Superheterodyne

9 Tubes with 10 TUBE PERFORMANCE

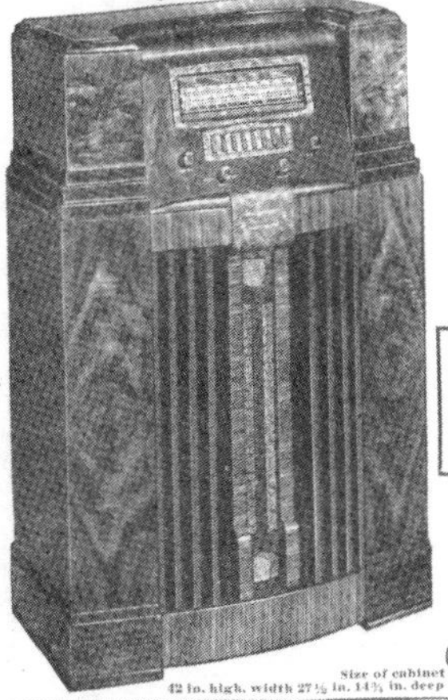

Size of cabinet 42 in. high, width 27½ in. 14½ in. deep

8 Button Magic Keyboard AUTOMATIC ELECTRIC TUNING

Built-In Magic Antenna—Needs No Aerial or Ground

Television Sound Provision and Connection for Record Player

Offers outstanding features that makes this STEWART-WARNER today's greatest console radio, because with it your customer can get new thrills and more real enjoyment from broadcasts all over the world.

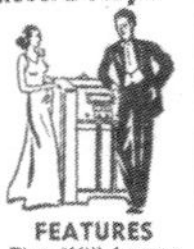

Foreign and American Short Wave—Standard BROADCASTS Coverage—540-1650 KC.—1625-5400 KC.—5400-18,000 KC. Includes Amateur, Aircraft, Ships-at-Sea, 2 Police Bands.

6 Station Magic Keyboard AUTOMATIC TUNING 8 push buttons for automatic tuning of favorite stations—6 for station selection, 1 for manual tuning, 1 for "phono-television." Wide vision dial for simplified Manual tuning.

No. 1RS 9A7 **46.58** NET IN LOTS OF 3

12 Inch DE LUXE Dynamic Speaker Finest, most true-toned speaker available. For wider tone range, more volume.

Tone Control—Tuning Control Band Switch 3 Position tone control for tone shading. Bass compensation. Full range tuning knob switch for any of the 3 Bands.

TODAY'S "BEST BUY" IN RADIO

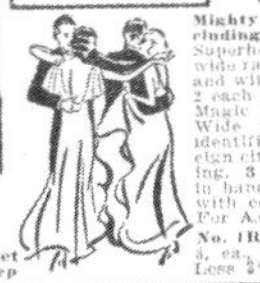

Mighty 10 tube performance with 9 tubes including rectifier and one dual purpose tube. Superheterodyne with **10 tuned circuits,** wide range band pass radio frequency stage, and with 1 each 12A8GT, 12Q7GT, 5W4G and 2 each 12SK7, 6J5GT, 6K6G, tubes. Built-in Magic Antenna, **no outside aerial needed.** Wide vision illuminated dial with band identification and names of important foreign cities shown for simplified manual tuning. **3 full tuning bands.** Cabinet finished in hand rubbed, polished American Walnut with center decoration of Carpathian Elm. For A.C. only. LIST, 94.60.

No. 1RS9A7. Dealer's each, **48.50.** 3, ea., **47.53.** Less 2%, net........................ **46.58**

FEATURES That Will Impress Your Customers

- 6 Station Automatic Electric Button Tuning
- Built-in Antenna. No Ground Needed
- 12-Inch P.M. Dynamic Speaker
- Lighted Large Wide Vision Dial
- Phono-Television Connections

Imperial Console Superheterodyne *AUTOMATIC* Record Changer RADIO-PHONOGRAPH

Luxurious Console Cabinet with Capacious Lower Shelf for Records

Offers more than double entertainment features—and double value for your customer's money. Brings quality of tone that satisfies everyone.

Offers Convenience of an automatic record changer, Phonograph and Radio, with beauty of a handsome console, that all admire.

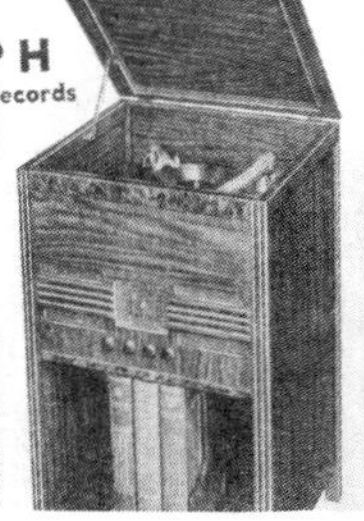

AUTOMATIC RECORD CHANGER PLAYS 10 and 12 INCH RECORDS

The very finest equipment is built into this Imperial. Powerful silent operating 78 RPM self-starting motor and an oversize turntable. **Record changer plays eight 10-inch or six 12-inch records automatically.** Tone arm of latest design is fitted with a **Phanstiehl permanent needle.** Has automatic stop, arm rest, needle cut, etc.—everything required for superb tone reproduction is built into this exceptional instrument.

Outstanding 5-tube Superheterodyne radio and phonograph combination in modern design console. Point for point here's a real buy for anyone who wants supreme value on a basis of true quality at low cost.

In Lots of 3, each, **46.55** NET

Unsurpassed for entertainment value—smart simplicity of modern style console. Plenty of space in capacious lower storage shelf for large library of phonograph records. Tone control allows selection of any degree of treble or bass emphasis as desired. Reproduction through the radio audio-system is remarkable for its brilliancy with rich depth and glowing reality—fidelity of both broadcasts or recorded programs developed in their original purity, providing the finest entertainment possible to obtain. Radio circuit, a 5-tube superheterodyne, uses the following super-sensitive matched tubes: **12SK7GT, 12SA7GT, 12SQ7G7, 50L6GT** and **35Z5GT.** Speaker unit is matched to the acoustics of the cabinet—**over-size**

AND THESE IMPORTANT **Features**

- *Lifetime Phanstiehl Phonograph*

1941 Catalog Page

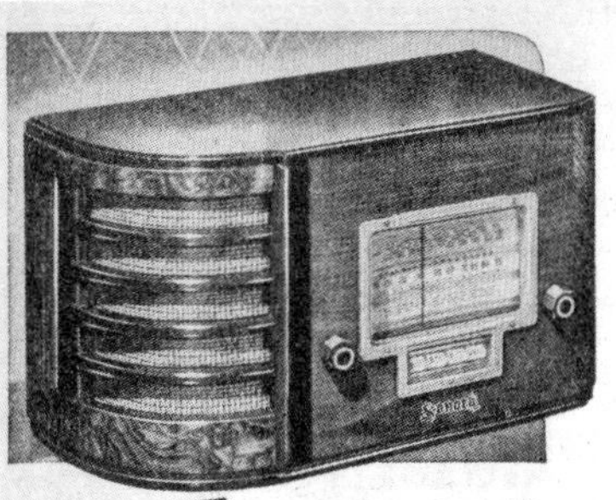

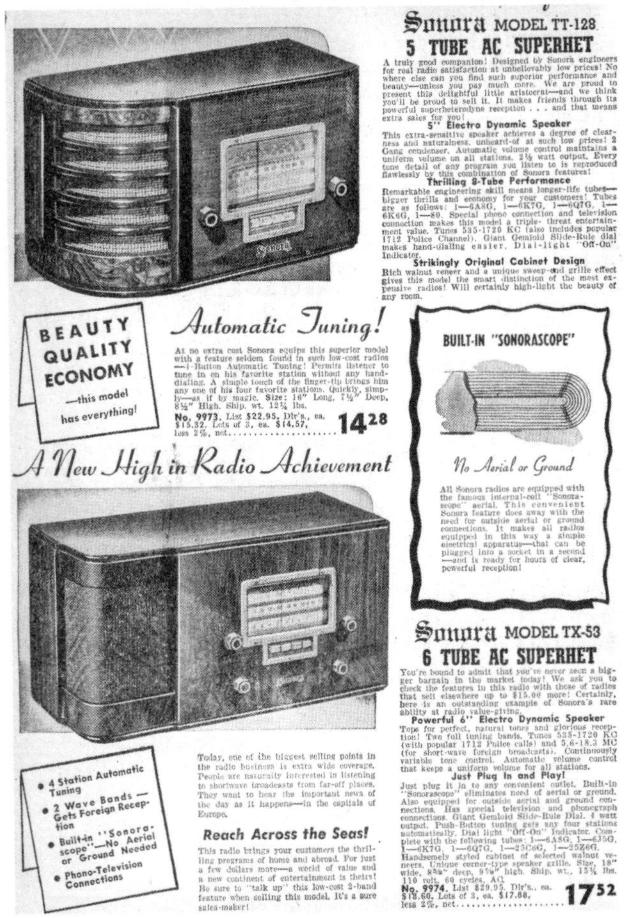

Sonora MODEL TT-128 5 TUBE AC SUPERHET

A truly good companion! Designed by Sonora engineers for real radio satisfaction at unbelievably low prices! No where else can you find such superior performance and beauty—unless you pay much more. We are proud to present this delightful little aristocrat—and we think you'll be proud to sell it. It makes friends through its powerful superheterodyne reception . . . and that means extra sales for you!

5" Electro Dynamic Speaker

This extra-sensitive speaker achieves a degree of clearness and naturalness, unheard-of at such low prices! 2 Gang condenser. Automatic volume control maintains a uniform volume on all stations. 2½ watt output. Every tone detail of any program you listen to is reproduced flawlessly by this combination of Sonora features!

Thrilling 8-Tube Performance

Remarkable engineering skill means longer-life tubes—bigger thrills and economy for your customers! Tubes are as follows: 1—6A8G, 1—6K7G, 1—6Q7G, 1—6K6G, 1—80. Special phono connection and television connection makes this model a triple- threat entertainment value. Tunes 535-1720 KC (also includes popular 1712 Police Channel). Giant Gemloid Slide-Rule dial makes hand-dialing easier. Dial-light "Off-On" Indicator.

Strikingly Original Cabinet Design

Rich walnut veneer and a unique sweep-end grille effect gives this model the smart distinction of the most expensive radios! Will certainly high-light the beauty of any room.

BEAUTY QUALITY ECONOMY —this model has everything!

Automatic Tuning!

At no extra cost Sonora equips this superior model with a feature seldom found in such low-cost radios—4-Button Automatic Tuning! Permits listener to tune in on his favorite station without any hand-dialing. A simple touch of the finger-tip brings him any one of his four favorite stations. Quickly, simply—as if by magic. **Size:** 16" Long, 7½" Deep, 8½" High. Ship. wt. 12¼ lbs.

No. 9973. List $22.95, Dlr's., ea. $15.32. Lots of 3, ea. $14.57, less 2%, net........................ **14.28**

BUILT-IN "SONORASCOPE"

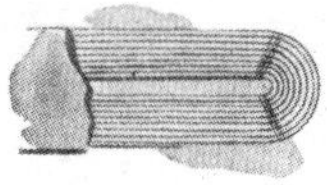

No Aerial or Ground

All Sonora radios are equipped with the famous internal-coil "Sonorascope" aerial. This convenient Sonora feature does away with the need for outside aerial or ground connections. It makes all radios equipped in this way a simple electrical apparatus—that can be plugged into a socket in a second—and is ready for hours of clear, powerful reception!

A New High in Radio Achievement

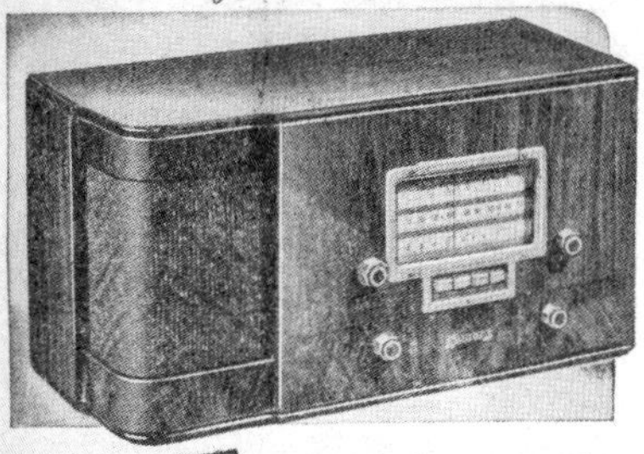

- 4 Station Automatic Tuning
- 2 Wave Bands — Gets Foreign Reception
- Built-in "Sonorascope"—No Aerial or Ground Needed
- Phono-Television Connections

Today, one of the biggest selling points in the radio business is extra wide coverage. People are naturally interested in listening to shortwave broadcasts from far-off places. They want to hear the important news of the day as it happens—in the capitals of Europe.

Reach Across the Seas!

This radio brings your customers the thrilling programs of home and abroad. For just a few dollars more—a world of value and a new continent of entertainment is theirs! Be sure to "talk up" this low-cost 2-band feature when selling this model. It's a sure sales-maker!

Sonora MODEL TX-53 6 TUBE AC SUPERHET

You're bound to admit that you've never seen a bigger bargain in the market today! We ask you to check the features in this radio with those of radios that sell elsewhere up to $15.00 more! Certainly, here is an outstanding example of Sonora's rare ability at radio value-giving.

Powerful 6" Electro Dynamic Speaker

Tops for perfect, natural tones and glorious reception! Two full tuning bands. Tunes 535-1720 KC (with popular 1712 Police calls) and 5.6-18.3 MC (for short-wave foreign broadcasts). Continuously variable tone control. Automatic volume control that keeps a uniform volume for all stations.

Just Plug in and Play!

Just plug it in to any convenient outlet. Built-in "Sonorascope" eliminates need of aerial or ground. Also equipped for outside aerial and ground connections. Has special television and phonograph connections. Giant Gemloid Slide-Rule Dial. 4 watt output. Push-Button tuning gets any four stations automatically. Dial light "Off-On" Indicator. Complete with the following tubes: 1—6A8G, 1—6J5G, 1—6K7G, 1—6Q7G, 1—25C6G, 1—25Z6G. Handsomely styled cabinet of selected walnut veneers. Unique corner-type speaker grille. Size, 18" wide, 8⅝" deep, 9⅞" high. Ship. wt., 15¾ lbs. 110 volt, 60 cycles, AC.

No. 9974. List $29.95, Dlr's., ea. $18.60. Lots of 3, ea. $17.88, less 2%, net........................ **17.52**

1941 Catalog Page

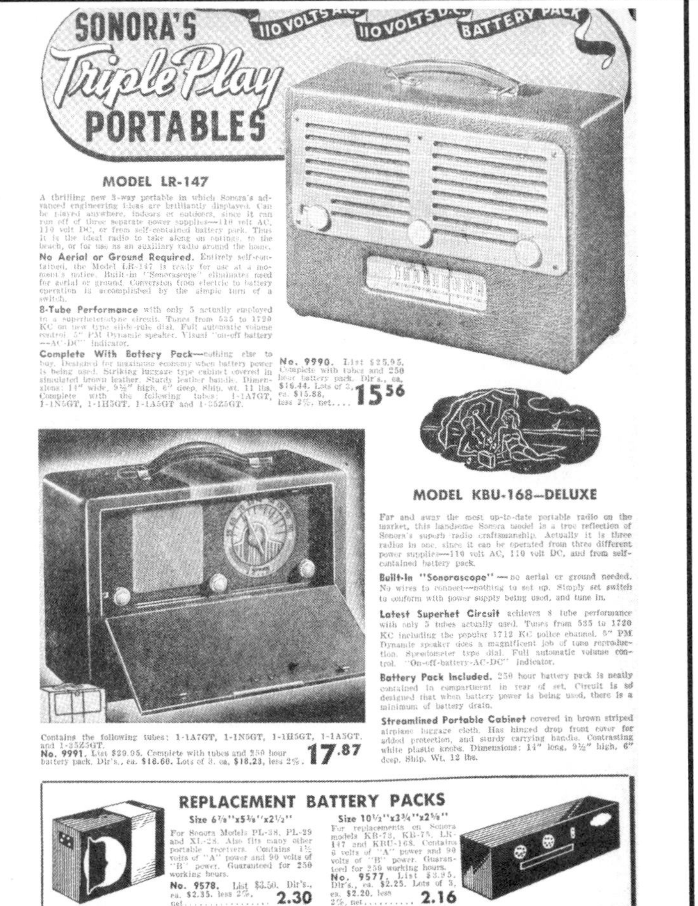

SONORA'S *Triple Play* PORTABLES

110 VOLTS A.C. 110 VOLTS D.C. BATTERY PACK

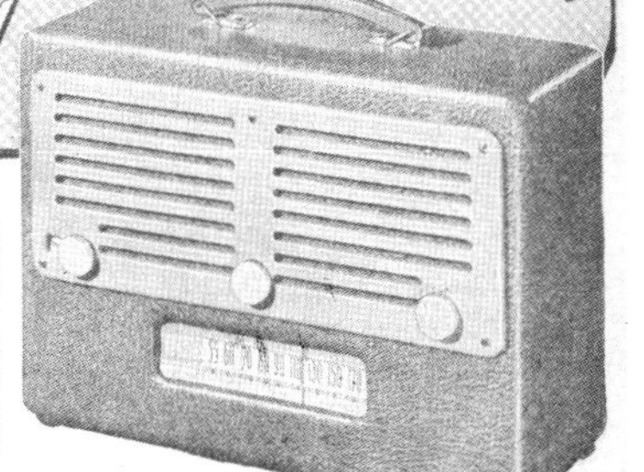

MODEL LR-147

A thrilling new 3-way portable in which Sonora's advanced engineering ideas are brilliantly displayed. Can be played anywhere, indoors or outdoors, since it can run off of three separate power supplies—110 volt AC, 110 volt DC, or from self-contained battery pack. Thus it is the ideal radio to take along on outings, to the beach, or for use as an auxiliary radio around the home.

No Aerial or Ground Required. Entirely self-contained, the Model LR-147 is ready for use at a moment's notice. Built-in "Sonorascope" eliminates need for aerial or ground. Conversion from electric to battery operation is accomplished by the simple turn of a switch.

8-Tube Performance with only 5 actually employed in a superheterodyne circuit. Tunes from 535 to 1720 KC on new type slide-rule dial. Full automatic volume control. 5" PM Dynamic speaker. Visual "on-off battery—AC-DC" indicator.

Complete With Battery Pack—nothing else to buy. Designed for maximum economy when battery power is being used. Striking luggage type cabinet covered in simulated brown leather. Sturdy leather handle. Dimensions: 14" wide, 9½" high, 6" deep. Ship. wt. 11 lbs. Complete with the following tubes: 1-1A7GT, 1-1N5GT, 1-1H5GT, 1-1A5GT and 1-35Z5GT.

No. 9990. List $25.95. Complete with tubes and 250 hour battery pack. Dlr's., ea. $16.44. Lots of 3, ea. $15.88, less 2%, net.... **15.56**

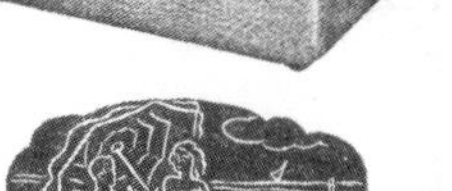

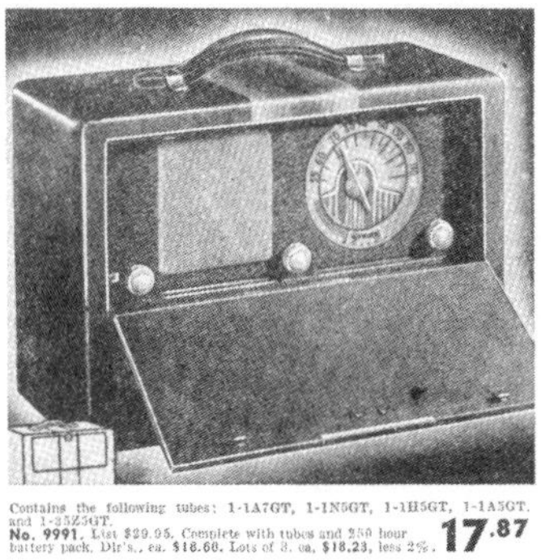

Contains the following tubes: 1-1A7GT, 1-1N5GT, 1-1H5GT, 1-1A5GT and 1-35Z5GT.

No. 9991. List $29.95. Complete with tubes and 250 hour battery pack. Dlr's., ea. $18.60. Lots of 3. ea. $18.23, less 2%. **17.87**

MODEL KBU-168—DELUXE

Far and away the most up-to-date portable radio on the market, this handsome Sonora model is a true reflection of Sonora's superb radio craftsmanship. Actually it is three radios in one, since it can be operated from three different power supplies—110 volt AC, 110 volt DC, and from self-contained battery pack.

Built-In "Sonorascope" — no aerial or ground needed. No wires to connect—nothing to set up. Simply set switch to conform with power supply being used, and tune in.

Latest Superhet Circuit achieves 8 tube performance with only 5 tubes actually used. Tunes from 535 to 1720 KC including the popular 1712 KC police channel. 5" PM Dynamic speaker does a magnificent job of tone reproduction. Speedometer type dial. Full automatic volume control. "On-off-battery-AC-DC" indicator.

Battery Pack Included. 250 hour battery pack is neatly contained in compartment in rear of set. Circuit is so designed that when battery power is being used, there is a minimum of battery drain.

Streamlined Portable Cabinet covered in brown striped airplane luggage cloth. Has hinged drop front cover for added protection, and sturdy carrying handle. Contrasting white plastic knobs. Dimensions: 14" long, 9½" high, 6" deep. Ship. Wt. 12 lbs.

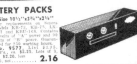

REPLACEMENT BATTERY PACKS

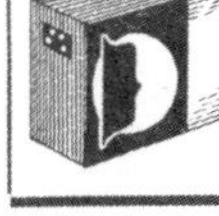

Size 6⅞"x5⅜"x2½"

For Sonora Models PL-38, PL-29 and XL-28. Also fits many other portable receivers. Contains 1½ volts of "A" power and 90 volts of "B" power. Guaranteed for 250 working hours.

No. 9578. List $3.50. Dlr's., ea. $2.35. less 2%, net................ **2.30**

Size 10½"x3¾"x2⅝"

For replacements on Sonora models KB-73, KB-75, LR-147 and KBU-168. Contains 6 volts of "A" power and 90 volts of "B" power. Guaranteed for 250 working hours.

No. 9577. List $3.95. Dlr's., ea. $2.25. Lots of 3, ea. $2.20. less 2%, net......... **2.16**

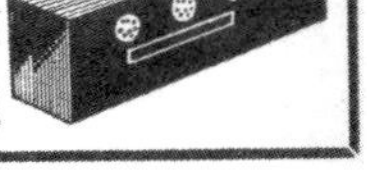

1941 Catalog Page

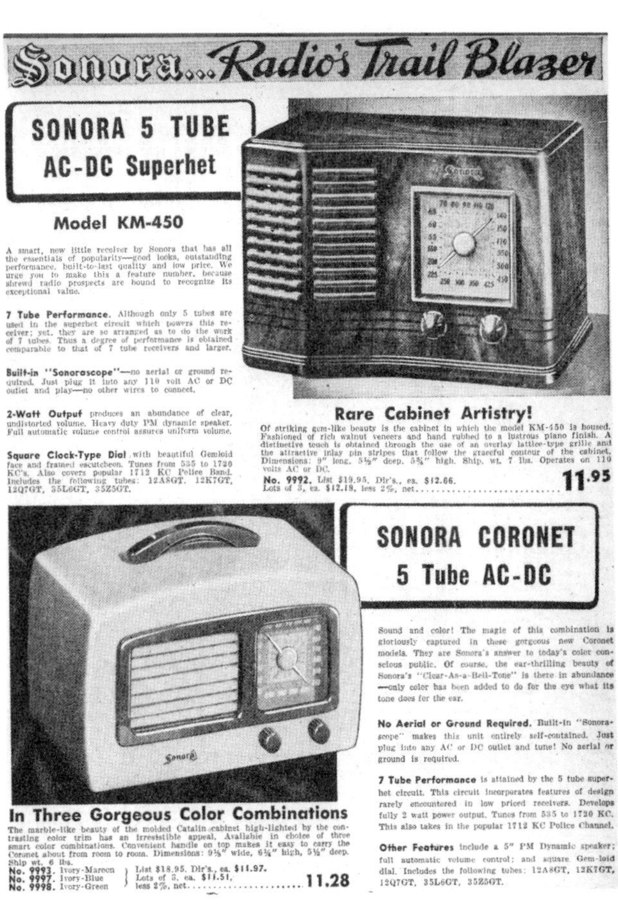

Sonora...*Radio's Trail Blazer*

SONORA 5 TUBE AC-DC Superhet

Model KM-450

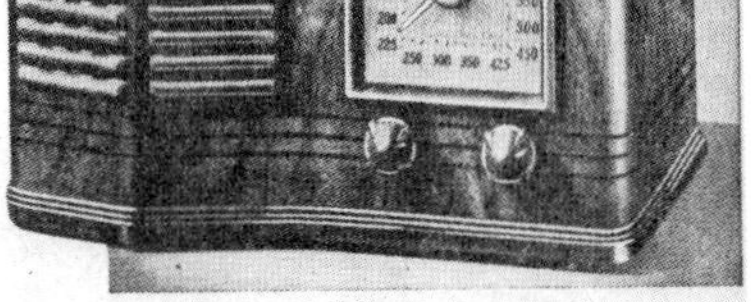

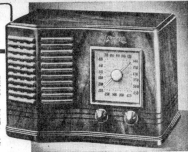

A smart, new little receiver by Sonora that has all the essentials of popularity—good looks, outstanding performance, built-to-last quality and low price. We urge you to make this a feature number, because shrewd radio prospects are bound to recognize its exceptional value.

7 Tube Performance. Although only 5 tubes are used in the superhet circuit which powers this receiver; yet, they are so arranged as to do the work of 7 tubes. Thus a degree of performance is obtained comparable to that of 7 tube receivers and larger.

Built-in "Sonorascope"—no aerial or ground required. Just plug it into any 110 volt AC or DC outlet and play—no other wires to connect.

2-Watt Output produces an abundance of clear, undistorted volume. Heavy duty PM dynamic speaker. Full automatic volume control assures uniform volume.

Square Clock-Type Dial with beautiful Gemloid face and framed escutcheon. Tunes from 535 to 1720 KC's. Also covers popular 1712 KC Police Band. Includes the following tubes: 12A8GT, 12K7GT, 12Q7GT, 35L6GT, 35Z5GT.

Rare Cabinet Artistry!

Of striking gem-like beauty is the cabinet in which the model KM-450 is housed. Fashioned of rich walnut veneers and hand rubbed to a lustrous piano finish. A distinctive touch is obtained through the use of an overlay lattice-type grille and the attractive inlay pin stripes that follow the graceful contour of the cabinet. Dimensions: 9" long, 5½" deep, 5¾" high. Ship. wt. 7 lbs. Operates on 110 volts AC or DC.

No. 9992. List $19.95. Dlr's., ea. $12.66. Lots of 3, ea. $12.19, less 2%, net........................ **11.95**

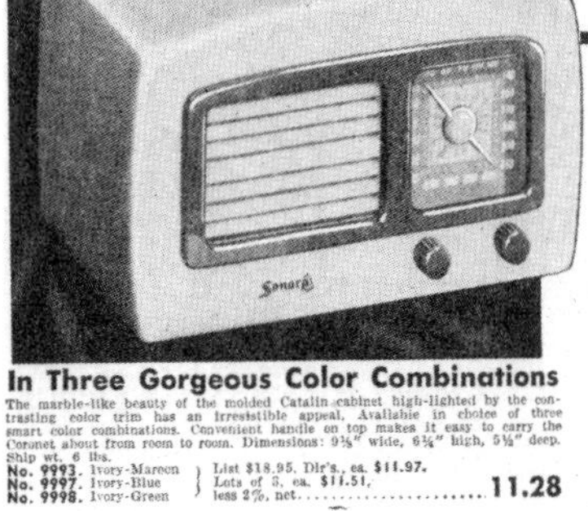

In Three Gorgeous Color Combinations

The marble-like beauty of the molded Catalin cabinet high-lighted by the contrasting color trim has an irresistible appeal. Available in choice of three smart color combinations. Convenient handle on top makes it easy to carry the Coronet about from room to room. Dimensions: 9⅛" wide, 6⅛" high, 5½" deep. Ship wt. 6 lbs.

No. 9993. Ivory-Maroon
No. 9997. Ivory-Blue
No. 9998. Ivory-Green
List $18.95. Dlr's., ea. $11.97. Lots of 3, ea. $11.51, less 2%, net........................ **11.28**

SONORA CORONET 5 Tube AC-DC

Sound and color! The magic of this combination is gloriously captured in these gorgeous new Coronet models. They are Sonora's answer to today's color conscious public. Of course, the ear-thrilling beauty of Sonora's "Clear-As-a-Bell-Tone" is there in abundance—only color has been added to do for the eye what its tone does for the ear.

No Aerial or Ground Required. Built-in "Sonorascope" makes this unit entirely self-contained. Just plug into any AC or DC outlet and tune! No aerial or ground is required.

7 Tube Performance is attained by the 5 tube superhet circuit. This circuit incorporates features of design rarely encountered in low priced receivers. Develops fully 2 watt power output. Tunes from 535 to 1720 KC. This also takes in the popular 1712 KC Police Channel.

Other Features include a 5" PM Dynamic speaker; full automatic volume control; and square Gem-loid dial. Includes the following tubes: 12A8GT, 12K7GT, 12Q7GT, 35L6GT, 35Z5GT.

1941 Catalog Page

PHILCO *Celebrates the*

15 millionth PHILCO RADIO

with the Greatest Values in its History!

PHILCO DEALERS FROM COAST TO COAST JOIN THIS GREAT CELEBRATION WITH

Special Jubilee Offers!

America has bought 15 million Philco radios . . . a proud record in the history of American industry . . . a mark of public preference not even remotely approached in its field.

And now, Philco celebrates! Throughout the nation, with Special Jubilee Offers, dealers present the greatest radio and radio-phonograph values Philco has ever produced . . . the crowning achievement of the skill and experience that have built 15 million Philco radios.

Spectacular Philco inventions have created a brand-new radio circuit that doubles selectivity, reduces noise and interference by 5 to 1 and makes overseas reception 5 times stronger, clearer and easier to tune. And, in addition, the 1941 Philco gives you more tubes for the money, bigger, finer speakers and an exquisite variety of lovely new cabinet designs.

In the new Philco *Photo-Electric Radio-Phonograph*, a new miracle of radio science plays any record on a *BEAM OF LIGHT!* Instead of a pointed steel needle that scrapes music from your records, a rounded, permanent jewel reflects the music from a tiny mirror to a photoelectric cell. No needles to change . . . record wear and surface noise reduced by 10 to 1 . . . glorious purity of tone . . . yes, new delights in the enjoyment of radio and recorded music!

Your Philco dealer presents these sensational 1941 Philco radios, radio-phonographs and auto radios now, in a great Celebration Sale! He offers liberal trade-in allowances, special easy payment terms and money-saving *Special Jubilee Offers*, all arranged for this event. It will pay you to visit him today!

See and hear the new PHILCO Radios, Radio-Phonographs and Auto Radios *from $9.95 to $395.*

1941 Magazine Ad

Give Them the Music You Want Them to Hear

TONIGHT—because someone was thoughtful—two little girls are holding a concert in their own room. They're listening to the music mother wants them to hear—and enjoying it!—on their new General Electric Radio-Phonograph!

Today—on the radio and on records—you will find much of the world's finest music—especially arranged for children. Why not give your children this wonderful opportunity to develop an appreciation for good music?

You can give them the G-E Radio-Phonograph shown for only $39.95*. It has excellent tone—touch-tuning radio keys and a sturdy, simple phonograph that even a child of three can operate.

Visit your General Electric Radio dealer tomorrow and ask especially to see Model J-678. You'll *always* be glad you bought a General Electric.

GET A G-E! YOU CAN'T BUY A BETTER RADIO!

Radio, as you know it today, was born in the General Electric "House of Magic"—it grew up there! Every set of every make contains fundamental features for which General Electric is responsible. But—only in a General Electric—can you get *all* the benefits of *all* the great advancements made to radio by General Electric.

The Blue Ribbon Prize Winner. An AC-DC superheterodyne with excellent tone and automatic volume control. Encased in handsome mahogany plastic cabinet that won the top award for styling in the nation-wide Modern Plastics contest. Model L-500. **ONLY $9.95***

A GENERAL ELECTRIC... THAT'S THE KIND WE USE ON OUR SQUAD CARS!

New—and Different! A light-weight carry-about radio—designed for action—in a small camera. You can take it with you wherever you go! Goes into battery. Remarkable tone! Case included in simulated leather with colorful plastic trim. Model J8-410. **Only $19.95***

Ideal for a Child's Room! A powerful AC-DC superheterodyne in a streamlined cabinet of walnut plastic. Built-in Beam-a-scope (No aerial—no ground) Remarkable selectivity. Wonderful tone. Model J-602.

All prices subject to change without notice and may vary in different localities. See your G-E dealer. General Electric Company, Bridgeport, Conn.

GENERAL ⊕ ELECTRIC *Golden Tone Radio*

1941 Magazine Ad

The *Ideal Gift*
IN A BETTER WORLD

Emerson Compact Model 502. (at right) AC-DC. Highly powered chassis with enclosed Super Loop and "Miracle Tone" speaker. New tube developments.

Emerson Portable Model 505. "3-Way" operation—battery, AC-DC. Amazing power for near and far-off broadcasts. Light weight and weather-resisting.

Emerson Phonoradio Model 506. Automatic record changer, aerial-powered radio with enclosed Super Loop and "Miracle Tone" Speaker. New plastic and metal features.

Emerson Pocket Radio Model 508. Self-powered—no outside wires. Light weight, unbreakable case. New tube developments. Fits easily in your pocket.

The New Post-War
Emerson Radio
and Television

Better Style, Tone, Performance and Value

A "Christmas Package" of enduring charm—a beautiful QUALITY instrument that will grace any home or office . . . the SPECIALIZED engineering product of the World's Largest Maker of Small Radio — truly the IDEAL Gift!

With THREE TIMES THE POWER OF PRE-WAR radio - with new construction, operating and reception improvements born of vast radio-radar-electronic research—with many other exclusive features—the post-war Emerson Radios and Phonoradios are "new world" in every respect.

The descriptions and prices of the models shown are only approximate. You may be sure that the set of your choice will surpass even those high standards which have gratified millions of Emerson Radio owners in the past. There are models for every purpose and every purse.

ORDER NOW

For early delivery of the sets of your selection, see your nearest Emerson Radio dealer and ORDER NOW.

EMERSON RADIO AND PHONOGRAPH CORPORATION, NEW YORK 11, N. Y.

WORLD'S LARGEST MAKER OF SMALL RADIO

1945 Magazine Ad

Bendix Radio
THE REAL VOICE OF

Beautiful

beyond belief in Tone and Styling

performance fine as the performer . . . cabinets distinguished as the best of heirloom furniture—these are only the highlights of the advantages coming to you when you brighten your home with a Bendix Radio. For each set—from the intricately new plastic table models to the finely-fashioned F.M. console combinations—is endowed with qualities found nowhere else. Better dealers everywhere will gladly demonstrate the welcome difference between a Bendix Radio and any radio you have heard or seen before.

BENDIX RADIO BENDIX RADIO DIVISION
BENDIX AVIATION CORPORATION, BALTIMORE 4, MARYLAND

PRODUCT OF Bendix AVIATION CORPORATION

1946 Magazine Ad

74

TABLE MODEL RADIOS

Order Yours on Time Payments

Graceful *Aircastle Battery Radio

6-Tube Performance—Precision Built

- Richly grained Walnut finish cabinet
- 4 tubes, two are dual purpose, giving 6-tube performance for clear reception
- Easily converted to an electric set
- 6-inch permanent magnet speaker

E **Modernly styled,** gracefully curved lines in a battery radio that makes a charming addition to every room. Built and engineered by skilled technicians. Has selective superheterodyne circuit and improved transformers for greater selectivity and sensitivity. Convertible to an electric set with battery eliminator No. A76H 1390 shown below. Hear your favorite music, news and market broadcasts in full clear tones on standard band 540 to 1725 KCs. Dimensions: 18x10x9 in.

B76H 263. With 1000-hr. Battery. (41 lbs.) **35.95**
B76H 264. Less battery. Shpg. wt. 16 lbs. **31.50**

RCA Victor Farm Battery Radio

Model 55F—With the "Golden Throat"

F Now you get ultra smart styling plus sparkling clear reception in a farm battery radio. RCA Victor 3-way acoustical system means unusual purity of tone.

7-tube performance—5 RCA Victor low drain tubes, two of the tubes are dual purpose. Operates on a single battery pack, convertible to 110-120 volt AC operation with the "Electro" battery eliminator shown below.

Improved reception with the automatic volume control, rubber-mounted tuning condenser, RCA teletube Radio-Frequency Amplification, selective superheterodyne circuit using 7 tuned circuits and powerful 5-inch permanent magnet dynamic speaker. Moisture-proof coils. Dimensions: 9¼ in. high, 18 in. wide, and 10½ in. deep. Uses RCA Farm Battery Pack No. VSO22 1½ volt A battery, 90 Volts. B Battery.

B76H 253. Including Batteries. Postpaid........ **48.15**
Selling price west of Rockies. Postpaid..... .. **50.25**
B76H 623. Less battery. Postpaid. **42.05**
Selling price West of Rockies. Postpaid......... **44.15**
A76H 1495. RCA Battery pack No. VSO22 for Model 55F—1½ volt A, 90 Volt B Battery. Postpaid..... **6.10**

*Aircastle Portable

6-Tube Performance AC - DC and Battery

G Handsome two-tone simulated leather portable that automatically plays on electric power or batteries. No switch is needed. Gives 6-tube performance—4-tubes plus rectifier—two tubes serve dual purpose.

All the most advanced engineering features are included; automatic volume control, large 5-in. permanent magnet dynamic speaker, built-in loop antenna, selective superheterodyne circuit, easy tuning, and all your favorite stations on the standard 540 to 1750 KCs tuning range. Dimensions: 8¾ in. high, 12 in. wide, and 6¼ in. deep. Operates on 110-120 volts, 50-60 cycles, AC and DC or requires two 45 volt B batteries and two 4½ volt A batteries. Batteries will last 250 hours.

B76H 701. Complete with batteries. Sh. wt. 7 lbs.**30.95**
B76H 700. Without batteries. Sh. wt. 10 lbs..... **27.95**

Battery Radio Electric Converter

Convert your battery radio into an electric one—economically, easily. Uses 110-120 volts AC and DC. Operates 4, 5 or 6-tube radios using 1½ volt A with 2 B batteries or a 1½ volt A in an AB pack. Simply plug in.
A76H 1390. Wt. 4 lbs...**14.69**

1946
Catalog
Page

E
Aircastle
Radio
35.95

F
RCA Victor
Radio
48.15

SPIEGEL • 11

G
Portable
Radio
30.95

SIX OF OUR FINEST

For Your Own Special Needs and Taste—

[A] Ivory Plastic **21.95**

Also in Walnut 19.95

[B] Ivory Finish **27.30**

Also in Walnut 25.40

[C] Walnut Veneer **33.45**

1946 Catalog Page

Gleaming Ivory or Walnut Plastic

6-Tube Performance—AIRCASTLE AC and DC Radio
- 4 Tubes plus one rectifier—2 tubes serve dual purpose—giving 6-tube performance
- Latest type 5-inch permanent magnet speaker
- Large built-in antenna, automatic volume control
- Illuminated slide-rule dial, with Lucite pointer
- Standard Tuning Range of 540 to 1720 KCs

[A] **Aircastle's distinctly modern,** full-size table model radio, not a miniature—at a remarkably low price. This radio is designed and engineered with the latest electronic developments. All above features generally found only in higher priced sets. Has selective superheterodyne circuit, and iron core I.F. transformer for greater selectivity and sensitivity.

Dimensions : 11½ in. wide, 8 in. high and 6 in. deep. Operates on 110-120 volts. AC and DC.

B76H 325. Ivory Plastic. Shpg. wt. 14 lbs......**21.95**
B76H 300. Walnut Plastic. Shpg. wt. 14 lbs......**19.95**

Glowing Splendor in Modern Plastic

7-Tube Performance RCA Victor Models 56X and 56X2
- With the "Golden Throat" 3-way acoustical system
- 5 tubes plus one rectifier; 2 tubes serve dual purpose, giving 7-tube performance
- 5-inch Super-sensitive electro-dynamic speaker
- Two-point tone control, automatic volume control
- Extra-large built-in magic loop antenna
- Standard Tuning Range 540 to 1620 KCs

[B] **Wonderful value** in the low priced radio field. RCA Victor with the "Golden Throat" acoustical system means magnificent reception. Amazing power for distance and for bringing true-to-life performance. The large plastic knobs provide hair-breadth tuning for your favorite programs. Truly an aristocrat among table model radios! Cabinet dimensions: 12 in. wide, 7¾ in. high, and 6¾ in. deep. Operates on 110-120 volts, 50-60 cycle, AC and DC. Postpaid.

B76H 257. Ivory finish. Model 56X2.........**27.30**
Selling Price West of Rockies. Postpaid........**28.65**
B76H 258. Walnut finish. Model 56X. Postpaid....**25.40**
Selling Price West of Rockies. Postpaid........**26.70**

*Aircastle Modern Walnut Veneer

7-Tube Performance—2-Band, AC and DC Radio
- 5 tubes plus one rectifier—2 tubes have dual purpose—giving 7-tube performance
- 5-inch permanent magnet dynamic speaker
- Variable tone control; easy tuning dial

[C] **Designed for a handsome piece** of furniture in your home. This Aircastle is as fine an instrument as you'll find anywhere for the money. All above features plus superheterodyne circuit, automatic volume control, large built-in loop antenna; and additional aerial connection. Standard tuning range brings your favorite programs on 540-1730 KCs, foreign short wave 5.4 to 18.3 MC. Size: 14½x9½x7½ inches. Operates on 110-120 volts, AC and DC.

B76H 331. Walnut Veneer. Shpg. wt. 15 lbs...**33.45**
*T.M.Reg. U. S. Pat. Off.

12-Foot House Mast Aerial

4.50
12 ft.

Finest quality telescoping aerial will improve foreign and home reception. Mounts on roof, soil pipe or at an angle from window. Built-in lightning arrestor and 60-ft. lead-in wire and mounting accessories.

A76H 1396. With instructions. Sh. wt. 5 lbs......**4.50**

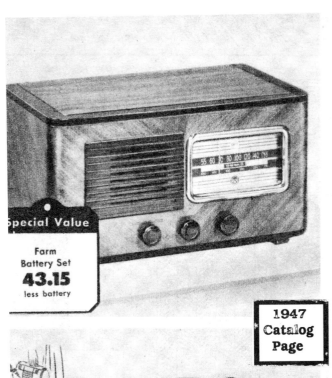

Special Value

Farm
Battery Set
43.15
less battery

**1947
Catalog
Page**

Light to carry

It's Tops!

3-Way
Portable
59.95
less Battery

Price Slashed

2-Way
Talking System
33.95

RCA VICTOR
BATTERY AND PORTABLE RADIOS

RCA Victor Precision-Built Farm Battery Radio

Highly polished Walnut cabinet plus purity of tone that only the "Golden Throat" can give . . . all engineered and built into this improved table model radio. Easy to convert to electricity.

Operates on a Single Battery Pack, convertible to 110-120 volt AC operation with battery eliminator 76K 1390 below. Handsome cabinet completely house battery. Battery-saver switch provides instant adjustment—cuts cost. On-off indicator. 5-low drain tubes combine excellent performance with economy of operation, 2 tubes are dual purpose, give 7 tube performance.

Wonderful performance on battery or power-line operation due to automatic volume control, rubber-mounted tuning condenser, RCA teletube Radio-Frequency Amplification, selective superheterodyne circuit using seven tuned circuits, and powerful 5-in. P.M. dynamic speaker. Tunes on the Standard Broadcast band of 540-1720 KCs and police calls. Uses battery pack below.

Dimensions: 9¼ in. high, 18 in. wide, 10½ in. deep.
B76K 618. Without batteries. We pay exp. charges.... **43.15**
Price West of Rockies. **45.30**
B76K 269. With batteries. We pay frt. charge. **50.65**
Price West of Rockies. **53.20**
A76K 1498. Battery pack VS022. 1½ volt A, 90 volts B. Postpd. **7.50**

Electro Battery Eliminator (Not shown) converts your battery radio to use 110-120 volts, 60 cycle, AC. Operates 4, 5, or 6 tube radios using 1½ volt A with 2 B batteries or a 1½ volt A battery in an A-B pack. Easy to install. Simply plug in.
A76K 1390. Shipping weight 4 lbs. **14.69**

The Globe Trotter — RCA Victor 3-Way Portable

Ideal traveling companion—plays at home, too. The new portable radio handsomely finished in aluminum and Maroon plastic case. It's attractively priced—exclusively designed—the latest word in style and performance. Exquisite tone through the exclusive qualities of the "Golden Throat." It's easy to carry—with a plastic handle molded to fit the hand. Easy tuning with gold color thumb wheels for station selection and volume control.

Full vision dial with iridescent pointer. Roll top switch turns set on or off with the flick of a finger. Use in trains and autos with specially designed removable loop antenna.

5-Tubes plus one rectifier—2 Tubes are dual purpose, giving 7-Tube performance. Small, compact, ultra modern portable offers exceptional performance at modest cost. Tunes on standard broadcast range of 540 to 1600 KCs.

Features extra selective tuning, automatic volume control, built-in Magic Loop antenna, highly sensitive P.M. dynamic speaker and selective superheterodyne circuit with 3 gang tuning condensers for improved reception. Operates on 110-120 volts, 50 to 60 cycles, AC and DC or on the self-contained long-life battery pack. We pay express charges.
B76K 706M. Portable Model 66BX. Less Battery. **59.95**
Selling Price West of Rockies **62.95**
A76K 1384M. Battery Pack for above. Postpaid. **5.25**

2-Way Talking System

These two-way talking systems were precision built by Western Electric. They're super-sensitive—they'll catch the slightest sound. Just flip a switch to speak or to listen.

2 amplifier tubes and a rectifier are included. All steel constructed Larger master unit has volume control. Can be operated in the home, school, restaurants, offices, factories and farms for two-way conversations. Instructions and directions included.

Master Unit measures 13x11½x6¾ in. Small unit measures 5x5x3½ in. Operates on 110-120 volts AC and DC current. Three speakers can be added. Shipped by freight or express.
B76K 316M. 2-Way System, 100 ft. wire. Shpg. wt. 25 lbs. **33.95**

Radiolink Extra Speaker (Not shown) For use as additional speakers with talking system shown above or with any radio. Easy to install. Has volume control and shut off switch. Can be operated with radio speaker shut off. Attractive simulated leather covered plywood case. Measures 9 in. wide, 6 in. high, and 3¾ in. deep. Order wire cord below.
A76K 318M. Speaker. Shpg. wt. 3 lbs. **6.65**
A76K 317M. Wire Cord. Shpg. wt. 2 oz. . . . . .Per foot **2½¢**

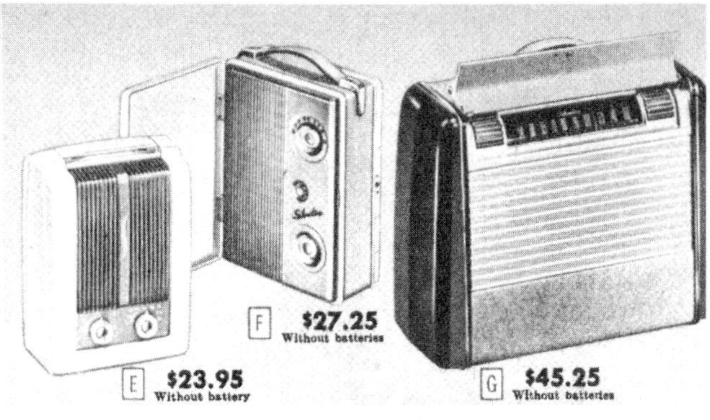

C $5.65   **D** $9.95

A Last Fall Price Was $47.75 **$44.95** Cash $4.50 Down

B $32.95

F $27.25 Without batteries

E $23.95 Without battery

G $45.25 Without batteries

SILVERTONES WITH FM . . extra listening fun!

A **Gray-green plastic cabinet.** Tunes 540–1600 KC broadcast *plus* 88–108 megacycle FM band. Built-in aerial for each band. Has 7 tubes plus rectifier; 5¼-in. speaker; variable tone control. Receives FM stations up to 30 miles distant (up to 50 miles with outdoor FM aerial). For 105–125 volt AC or DC. UL approved.
57 H 08021—Sizes 8⅝x14¼x8³⁄₁₆ inches. Shpg. wt. 12 lbs. Mailable........$44.95

B **Ivory-color plastic cabinet,** gold-color trim. Tunes broadcast and FM bands. Has 7 tubes plus rectifier; 4-in. speaker. For 105–125 volt, 50–60 cycle AC. Recommended only for local reception within 15 miles of an FM station.
57 H 08022—Size 7x11x6½ inches. Shipping weight 10 lbs. Mailable........$32.95

C 57 H 06706—**Outdoor FM Antenna.** Height 5 ft. Must be directed to station. Includes hardware, instructions. 50-ft. down-lead, base. Shpg. wt. 8 lbs......$5.65

D 57 H 06710—**Turnstile FM Antenna.** Height 7¾ ft. Receives well from all directions. Includes hardware, 50-ft. down-lead, base. Shpg. wt. 10 lbs.........$9.95

K Last Fall Price Was $22.65 NOW $19.95

H $9.95 Gray-Green

J $15.95   **L** $29.95

M $49.95

N $89.50   **P** $95.25

R $30.10   **T** $50.25   **W** $14.95

S $40.25

V

232 . . SEARS, ROEBUCK AND CO.

PORTABLE RADIOS . . for AC, DC or Batteries

E **Ivory-color plastic case.** Size 9¾x6¾x4 in. Built-in aerial; 4 tubes plus rectifier; 3½-in. speaker. For 117 volts AC or DC or for battery. UL approved.
57 H 09260—Without battery. Shpg. wt. 6 lbs. *Last Fall price was $24.95*......$23.95
57 HT 9261—Silvertone Personal Portable with battery. Shpg. wt. 7 lbs........... 26.34

F **Metal case,** 8⅜x6⅞x3½ in. Tan artificial leather finish. Ivory-color hinged plastic covers. Detachable shoulder strap. Built-in aerial; 4 tubes plus rectifier; 3½-in. speaker. For 105–125 volts AC or DC or for batteries. UL approved.
57 H 08260—Without batteries. Shpg. wt. 7 lbs. *Last Fall price was $28.75*......$27.25
57 HT 8261—De luxe Personal Portable with battery. Shpg. wt. 8 lbs 29.50

G **Extra-powerful!** Has 5 tubes plus rectifier; 5¼-in. speaker; built-in aerial. Aluminum case, painted to resemble tan leather. Brown plastic ends. Size 14x11x5⅝ in. Operates on 105–125 volts AC or DC or on Battery. UL approved.
57 H 08270—Without battery. Shpg. wt. 14 lbs. *Last Fall price was $50.25*......$45.25
57 HT 8271—Silvertone De luxe Portable with battery. Shpg. wt. 21 lbs 49.25

TABLE RADIOS . . add charm to any room

H **Lowest priced Silvertone!** Metal cabinet, 6½x5x4½ in. Three tubes plus rectifier; 4-in. speaker. Tunes 540–1600 KC. For 105–125 volt. 25–60 cycle AC or DC. UL approved. Recommended only if you have a local radio station in your own town.
57 H 08004—Ivory-color cabinet, gold-color trim. Shipping weight 5 pounds.....$11.95
57 H 08003—Gray-green cabinet. Shipping weight 5 pounds. Mailable......... 9.95

J **New low-priced Silvertone.** Brown plastic cabinet. Built-in aerial; 4 tubes plus rectifier; 4-in. speaker. Tunes 540–1600 KC. Size 8⅝x5¾x6 in. For 105–125 volts AC or DC. Recommended only if you have a local radio station in your town.
57 H 09000—Underwriters' Labs., Inc. approved. Shpg. wt. 7 lbs. Mailable...$15.95

K **Silvertone in smart plastic cabinet;** enclosed, molded back. Gray-green finish. Built-in aerial; 4 tubes (2 dual-purpose) plus rectifier; 4-in speaker. Tunes 540–1600 KC. Size 6¾x6¹⁄₁₆x11⅜ in. For 105–125 volt AC or DC
57 H 08005—Underwriters' Labs., Inc. approved. Shpg. wt. 8 lbs. Mailable..$19.95

L **Silvertone with Automatic Clock.** Wake up to music! Your program will come on at exactly the time you set. Built-in aerial; 4 tubes plus rectifier; 4x6-in. oval speaker. Tunes 540–1600 KC. Ivory-color plastic cabinet; enclosed, molded back. Gold-color mesh grille. Size 7⅝x12½x6¾ in. For 105–125 volt. 60 cycle AC.
57 H 08011—Underwriters' Labs. Inc. approved. Shpg. wt. 11 lbs. Mailable..$29.95

M **Hallicrafters S-38.** Tunes 540–1600 KC *plus* 3 bands of short wave from 1.6–32 megacycles. Has regular tuning plus fine tuning (Band Spread). Has 5 tubes plus rectifier; 5-in. speaker; jack for connecting headphones. Gray steel cabinet, 12⅞x 7⅞x8⅝ in. For 105–125 volts AC or DC.
57 H 07450—Underwriters' Labs., Inc. approved. Shpg. wt. 14 lbs...........$49.95

N **Hallicrafters S-53.** Tunes 540–1600 KC *plus* 4 short wave bands from 1.6–54.5 megacycles. Regular tuning plus fine tuning (Band Spread). Has 7 tubes plus rectifier; 5-in. speaker; jacks for headphones and record player. Two-tone gray steel cabinet, 12⅞x7⅞x6⅞ in. For 105–125 volt. 50–60 cycle AC.
57 H 07451—Underwriters' Labs., Inc. approved. Shpg. wt. 23 lbs...........$89.50

P **Hand-rubbed mahogany veneer cabinet.** More "reaching-power" on broadcast band and short wave than any other SILVERTONE ever! Tunes 540–1600 KC *plus* 4 short wave bands from 1.62–22.2 megacycles. "Climatized" to stand up anywhere from tropics to the arctic. Big 8-in. speaker; tone control. Input jack for record player. Size 17¼x10¾x13 in For 110–125–145–200 or 245 volts, 40–60 cycle AC.
57 H 09054—Underwriters' Labs., Inc. approved. Shpg. wt. 30 lbs...........$95.25

BATTERY OR HI-LINE RADIOS by Silvertone

SILVERTONE Battery Radios give you 4-way savings — 1. Low in price! 2. Easy on your batteries! 3. Battery included! 4. Operate on Hi-Line when it comes through! Radios (R). (S). (T) include Powr Shiftr (see page 237 for description). Radio (V) has built-in power supply for use on Hi-Line. All radios tune 540–1600 KC. Battery Packs fit inside radio cabinet. Includes Stratobeam Receptor to improve reception.

R **Gray-green plastic cabinet.** Has 4 tubes; 5¼-in. speaker. Size 9¼x15¼x7¾ in.
57HT8203—With 400-hour Battery Pack and Powr Shiftr. Shpg. wt. 26 lbs...$30.10

S **Walnut veneer cabinet.** Has 5 tubes; 5¼-in. speaker; tone control; thrift switch. Size 11x17¼x10⅞ in. Includes 1000-hour Battery Pack and Powr Shiftr.
57 HT 8222—Shpg. wt. 51 lbs. *Last Fall Price Was $57.70 (sold separately)*....$40.25

T **Radio-Phonograph.** Crystal pick up; long-life needle; spring-wound motor. Has 4 tubes (2 dual-purpose); 5¼-in. speaker. Walnut veneer cabinet, 10x15¾x13 in.
57 HT 8213—With 400-hour Battery Pack and Powr Shiftr. Shpg. wt. 42 lbs...$50.25

V **Mahogany veneer cabinet,** 11½x17½x11 in. Tunes 540–1600 KC *plus* 3 short wave bands (9.4–9.7, 11.6–12, 15–15.5 megacycles). Has 5 tubes plus rectifier; 5¼-in. speaker; tone control; thrift switch. For 105–125 volts, 60 cycle AC or for battery.
57 HT 8231—With 1000-hour battery. UL approved. Shpg. wt. 45 lbs........$60.25

W **Record Player** plays through your battery radio. Hand-wound motor; crystal pick up; long-life needle. Brown metal base. Size 12¼x11¾x5¼ inches.
57 H 08142—Shpg. wt. 12 lbs. *Last Fall Price Was $18.00*..................$14.95
your radio is not wired for record player, order correct Adapter below.
H 8146—For radios using tubes 1LA4 or 1LB4. Shipping weight 2 oz.......$1.00
H 8147—For radios using tubes 1A5, 1C5, 1G5 or 3Q5. Shpg. wt. 2 oz....... 1.00

ARVIN RADIOS

THE SATURDAY EVENING POST

Collier's

5-TUBE—NEW BEAUTIFUL CATALIN CABINETS

With Built-in Antenna - - Airplane Dial - - Superheterodyne Circuit
*Burgundy CATALIN with Onyx Trim
*Onyx CATALIN with Amber Trim

WHY
It Pays You
To Feature
ARVIN

Offer the radio that millions prefer . . . no radios are better built than Arvin. No radio except Arvin "clicks" so often with sure-fire hits of outstanding beauty. And only Arvin consistently produces year after year the **most** radio set f o r **least** money. Arvin adds prestige to your entire radio stock. Means MORE sales on every make and type you handle.

Eye-catching, luxurious beauty never before offered at even near a price so low. Will sweep the field in 1942. You'd expect that from Arvin—a name that practically every customer realizes must stand for the finest radio quality that money can buy. Your customer will hear performance surpassing anything that the average radio listener has ever known from a 5-tube compact set. Arvin-built superheterodyne circuit uses one each of low-drain 12SA7GT, 12SK7GT, 12SQ7GT, 50L6GT and 35Z5GT preferred type tubes and quadruple tuned I.F. amplifier. Built-in **Phantom-Scope** antenna eliminates need for other aerial, or for ground, so that set can be used anywhere. Input power—30 watts. Output power—2 watts. Resistance coupled Class "A" amplifier. Electrodynamic speaker. Covers the entire 540 to 1730 Kc. range of the American Standard Broadcast Band, and some police calls. Has airplane type dial. Pilot lamp. Catalin Knobs to Match grill. Size 8½x6¼x5¾ in. AC or DC.

No. GR532. BURGUNDY, onyx trim. LIST, 16.95.
Dealer's, each, 11.86. 3, each, 11.02. Less 2%, net..... **10⁸⁰**
No. GR532A. ONYX, amber trim. LIST, 16.95.
Dealer's Price, each, 11.86.
Lots of 3, each, 11.02.
Less 2%, net.................... **10⁸⁰**

6-TUBE ARVIN Superhet - - Plastic ON ALL FOUR SIDES

Has Sensational Phantom-Scope Built-in Aerial

This beautifully streamlined plastic radio presents as handsome a view from the back as from the front—ideal to use on tables that stand in the center of the room. Convenient carrying handle **makes moving easy.** Tubes include 2 dual-purpose tubes; make this radio equal in performance to the average 8-tube set. I. F. Amplifier provides sharp selectivity. Covers 540-1600KC for all standard American Broadcasts. Variable tone control. Magnificent reproduction over wide tone range insured by powerful "large area" permanent Magnet speaker. Phantom-scope built-in antenna affords greater receptive range. Needs no ground nor external aerial, but connection is provided for outside aerial if desired. Cabinet 12⅜x7¼x7 in. Large lighted airplane dial. Tubes: 1-12SA7, 2-12SK7, 1-12SQ7, 1-35L6GT, 1-35Z5GT, including rectifier. **For AC or DC.**

* Equipped with Dual Purpose Tubes - - Performs Like Many Sets with 2 More Tubes

No. GRAV722A. IVORY. LIST, 22.95.
Dealer's, each, 14.92. 3, each, 14.22.
Less 2%, net.................... **13⁹³**

No. GRAV722. WALNUT. LIST, 21.95.
Dealer's, each, 14.27. 3, each, 13.62.
Less 2%, net.................... **13³⁴**

The Back Is Also Streamlined

5-TUBE ARVIN Superheterodyne

With 2 Dual-Purpose Tubes

Using Dual Purpose Tubes This Set With Only 5 TUBES Actually Performs Like Many Sets With 7 TUBES!

Remarkable superheterodyne 5-tube circuit, of which 2 are dual purpose tubes, developing 7-tube performance. Covers full band of 540-1600KC.—all standard American Broadcasts. Quadruple tuned I. F. amplifier assures sharp selectivity, eliminates interference from "close" stations, as in ordinary radios. Built-in "hank-type" antenna—plays from any outlet. Connecting with outside aerial gives even greater distance. Electro-dynamic speaker. 5 tubes: 12SA7, 12SQ7, 12SK7, 50L6GT, and 35Z5GT, including rectifier. **Unbreakable cabinet** 7½ x 5½ x 4¾ in. **Lighted airplane dial. A.C. or D.C.**

No. GRAV524A. IVORY. LIST, 12.95.
Dealer's, each, 9.71. 3, each, 9.07.
Less 2%, net.................... **8⁸⁸**
No. GRAV524. WALNUT. LIST, 11.95.
Dealer's, each, 8.96. 3, each, 8.60.
Less 2%, net.................... **8⁴²**

4-TUBE "ARVINET" SUPERHETERODYNE

Unbreakable Ivory or Walnut Cabinet

6³² Walnut

1949 Catalog Page

Always a popular sales leader since introduced—now more than ever improved in performance and appearance, this smallest ARVIN Table Model attracts attention with its compact size; furnishes a new thrill with its surprising improved **4-tube superheterodyne performance.** A demonstration—and its exceedingly low price—draws easy and profitable sales. Weighs only 4 pounds; so tiny it can be tucked in a suitcase or carried in hand in its special suede carrying case.

This model performs like many larger sets. Uses one each 12SA7, 12SQ7, 50L6GT and 35F5GT tubes; resistance coupled class A Audio amplifier; **electro-dynamic speaker; automatic volume control; attached antenna.** Uses no ballast nor line cord resistor. Operates anywhere on 60-cycle AC or DC. Embossed bronze dial.

● Covers 540-1750Kc—standard American broadcasts — some police calls. Operates anywhere on 60 cycle AC or DC.

No. GRAV-422A. IVORY. LIST, 9.45.
Dealer's, each, 7.09.
3, each, 6.62. Less 2%, net........ **6⁴⁸**
No. GRAV-422. WALNUT. LIST, 8.95.
Dealer's, each, 6.71.
3, each, 6.45. Less 2%, net...... **6³²**

ZIPPER CASE

Well made from rubber-bonded fawn-and-brown suede. With 2 strap handles.

LIST, 1.00.
No. IRAV-1.
Each, 70c.
Less 2%, net.... **68c**

79

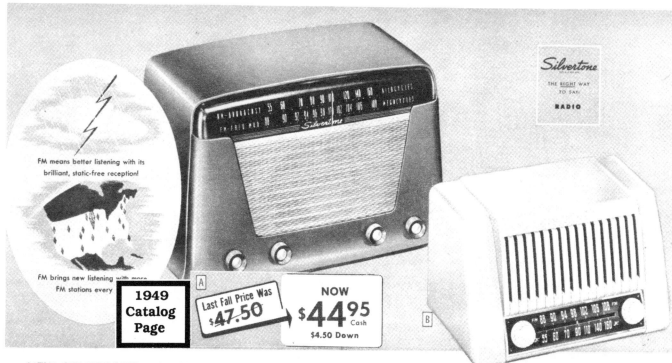

FM means better listening with its brilliant, static-free reception!

FM brings new listening with more FM stations every

1949 Catalog Page

Last Fall Price Was $47.50

NOW $**44**⁹⁵ Cash
$4.50 Down

NEW SILVERTONE with AM plus FM—enjoy rich-toned standard broadcasts . . thrilling FM with new, improved "reaching-power!"

[A] Now you can enjoy SILVERTONE's most popular AM-FM radio in a new version that gives you richer tone and greater power—all at an economy price! Yes, the newly engineered circuit gives you twice as much sensitivity as before! That means amazing new "reaching-power" . . . you can receive FM stations as far as 30 miles distant. If you use an outdoor FM aerial (see page 638), you can enjoy listening to FM stations as far as 50 miles away.
Tunes all the fun and entertainment on the 540-1600

KC broadcast band plus the static-free clarity of the 88-108 megacycle FM band. Separate built-in aerial for each band. Newly designed circuit uses 7 tubes plus selenium rectifier that gives longer life, greater power output. Variable tone control; 5¼-inch Alnico 5 speaker.
Handsome gray-green plastic cabinet. Size 8⅜x-14¼x8³⁄₁₆ in. Operates on 105-125 volts AC or DC. Approved by Underwriters' Labs., Inc.
57 E 08021—Shpg. wt. 12 lbs. Mailable........$44.95

NEW AM-FM RADIO
at a low, budget-fitting price

[B] Sleek, modern SILVERTONE Radio has ivory-color plastic cabinet, gold-color trim. Built-in aerial for broadcast band. Aerial for 88-108 megacycle FM band is attached to rear of set. Has 7 tubes plus rectifier; 4-in. speaker. Operates on 105-125 volt, 50-60 cycle AC. Size 7x11x6½ inches. Recommended only for local reception within 15 miles of an FM station.
$**32**⁹⁵ Cash
$3.50 Down
57 E 08022—Shpg. wt. 10 lbs. Mailable......$32.95

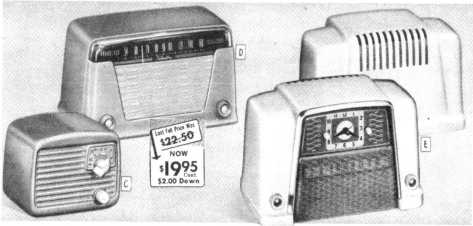

Last Fall Price Was $22.50
NOW $**19**⁹⁵ Cash
$2.00 Down

SILVERTONE TABLE RADIOS . .
give low-priced entertainment

[C] **Lowest priced Silvertone!** Extra-compact, too . . . only 6½x5x4½ in. Nonbreakable metal cabinet. Three tubes plus rectifier; 4-in. speaker. Tunes 540-1600 KC. For 105-125 volt, 25-60 cycle AC or DC. Underwriters' Labs., Inc. approved. Recommended only if you have a local radio station in your own town. Shpg. wt. 5 lbs. Mailable.
Gray-green $**9**⁹⁵
57 E 08004—Ivory-color finish, gold color trim.......$11.95
57 E 08003—Gray-green finish....................9.95

[D] **Rich-toned, powerful Table Radio.** Smart plastic cabinet with enclosed, molded back; gray-green finish. Has 4 tubes (2 dual-purpose) plus rectifier; 4-in. speaker; built-in aerial. Handsome aluminum grille. Tunes 540-1600 KC. Size 6½x6¹⁄₁₆x11⅞ in. For 105-125 volt, 25-60 cycle AC or DC. Underwriters' Labs., Inc. approved.
57 E 08005—Shpg. wt. 8 lbs. Mailable............$19.95

SILVERTONE CLOCK RADIO . .
turns itself on automatically!

[E] **Wake up to music!** Set the electric clock for any program, day or night . . . then tune in the station you want. The radio will turn itself on at exactly that time! Let this SILVERTONE remember your favorite comedy and news programs for you. Distinctive ivory-color plastic cabinet has an enclosed, molded back; contrasting gold-color mesh grille. Size 7⅝x12½x6¾ inches.
$**29**⁹⁵ Cash
$3.00 Down
Superheterodyne radio gives 6-tube performance from 4 tubes (2 dual-purpose) plus rectifier; 4x6-inch oval speaker; built-in aerial. "Ghost dial" appears only when set is turned on. Tunes 540-1600 KC. Operates on 105-125 volt, 60 cycle AC only. For your protection, approved by Underwriters' Laboratories, Inc.
57 E 08011—Shpg. wt. 11 lbs. Mailable...........$29.95

Use Sears Easy Terms . . . see inside back cover

NEW TABLE SILVERTONE

Four tubes(2 dual-purpose type) plus rectifier

Built-in loop aerial . . . no unsightly hanging wires

[F] **Sears lowest price in years** for a set of this quality! Handsome brown plastic cabinet, gold-color trim. Harmonizes with any furnishings. Has full-toned 4-inch speaker. Size 8⅝x5¾x6 inches. Operates on 105-125 volt, 25-60 cycle AC or DC. UL approved. Recommended only if you have a local radio station in your own town.
$**15**⁹⁵
57 E 09000—Shpg. wt. 7 lbs. Mailable..$15.95

c 627 .. TABLE MODEL RADIOS

AIRCASTLES—BIG VALUES—LOW PRICES

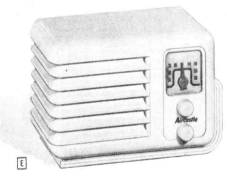

E BARGAIN-PRICED AIRCASTLE MIDGET
Only **8.98**
• Extra-sturdy! Compact! Lightweight!
• Modern streamlined plastic cabinet
• Ideal for bedroom, kitchen or den
in Walnut
HANDIEST TABLE MODEL IMAGINABLE! Use it in any room in the home. Just pick it up and plug it in whenever and wherever it suits you. Unites high fidelity tone with a budget-price that's hard to beat. Delights your pocketbook as well as your ear.
SPARKLING RECEPTION! Powerful PM. dynamic speaker—with heavy Alnico #5 magnet reproduces tone with utmost fidelity. Gives clear-as-a-bell 4-tube performance on 3 tubes (1 dual purpose) plus rectifier.
PRECISION-ENGINEERED. Sensitive superheterodyne circuit. Attached antenna. Easy-to-read airplane type dial. Lovely-to-look at Walnut or Ivory finish. Standard Range: 540-1700 kilocycles, 105-125 volts, 50-60 cycles, AC or DC. (Not recommended for use over 50 miles from a station.) Mailable from Chicago.
A76R 371. (E) Walnut. 7½x4¾x5¼ in. high. (7 lbs.) 8.98
A76R 372. (E) Ivory. 7½x4¾x5¼ in. high. (7 lbs.) 9.98

3 WAYS TO ORDER YOUR AIRCASTLE
• CASH   • 30-DAY CHARGE
• MONTHLY PAYMENT PLAN

A HIGH QUALITY AIRCASTLE PRICED LOW
• Richer tone with P.M. dynamic speaker   **15.95**
• Distance-getting built-in loop antenna
• Extra sensitive superheterodyne circuit   In Walnut
PACKS A WEALTH OF LISTENING PLEASURE into an up-to-date radio priced for the budget-minded. Hear it once, and you'll say there's no greater bargain in good listening.
POWERFUL 6-TUBE PERFORMANCE on 4 advanced-type tubes (2 dual purpose) plus rectifier. Large 5-in. Permanent Magnet speaker reproduces tone with superb fidelity. With heavy Alnico No. 5 magnet—2½ times more powerful than old types.
INTERFERENCE-FREE RECEPTION is assured by the ultra-modern superheterodyne circuit. Large, built-in loop antenna pulls in distant stations. Eliminates outside wiring problems.
NO FADING AND BLASTING! Automatic volume control puts an end to these old fashioned annoyances. Lightweight aluminum chassis. Airplane-type dial. Ivory or Walnut finished plastic cabinet. Standard Range: 540-1700 kilocycles. 105-125 volts, 50-60 cycles, AC or DC. Mailable from Chicago.
A76R 362. (A) Walnut. 12x6¼x7 in. high. (12 lbs.) 15.95
A76R 363. (A) Ivory. 12x6¼x7 in. high. (12 lbs.) 16.95

C SUPERB FM/AM PLASTIC TABLE MODEL
• Brilliant tone quality of modern FM   **34.95**
• Plus the richness of time-proven AM
• Remarkable built-in serial system   or $5 monthly
DOUBLE YOUR LISTENING ENJOYMENT with this all-new Aircastle. Thrill to the glorious world of radio sound that FM (Frequency Modulation) alone can bring you. FM eliminates the "blind spots" from listening—reproduces sound with true-to-life richness. Give you static-free reception even under unfavorable atmospheric conditions. Amazing table model doubles your listening choices. Allows you to tune in on the wonderful new FM programs or your old favorites on improved AM (Standard Broadcast.)
BIG 8-TUBE PERFORMANCE is surprising on budget-priced set. Has 7 tubes (1 dual purpose) plus rectifier. Powerful 4-in. speaker with heavy Alnico No. 5 magnet. Advanced design superheterodyne circuit. Built-in FM/AM aerial system.
RAZOR-SHARP TUNING made easy by clear-view tuning panel. Automatic volume control eliminates blasting and fading. Beautiful walnut-finished plastic cabinet. AM Band: 550-1600 kilocycles. FM Band: 88-108 megacycles. 105-125 volts, 60 cycles, AC only. Mailable from Chicago.
A76R 358. (C) AM/FM. 11¼x6½x7 in. high. (14 lbs.) 34.95

B NEW POWER-PACKED TABLE MODEL
Only
• Clear-as-a-bell tone reproduction   **22.5**
• Pulls in distant stations with ease
• Rosewood-finished plastic cabinet
COMBINES SUPERB RECEPTION AND BEAUTY in a thr priced Aircastle guaranteed to please the most discriminati listener. Lovely rosewood-finished cabinet enhances the attra tiveness of any room. High fidelity tone reproduction of the lar 5-in. speaker—equipped with modern Alnico No. 5 magnet will amaze you with it's life-like clarity.
SUPER-SENSITIVE SUPERHETERODYNE CIRCUIT features high-efficiency tubes plus advanced-type rectifier. Economic ly delivers 6-tube performance—one tube is dual purpo Long-range reception with built-in loop antenna.
AUTOMATIC VOLUME CONTROL increases your listenii pleasure. Does away with blasting and fading. Rugged plas cabinet designed to protect working parts from shock. Easy-read slide rule dial for razor-sharp station selection. Standa Broadcast Range: 550-1700 kilocycles. Operates on 110-1 volts, 50-60 cycles, AC or DC. Mailable from Chicago.
A76R 373. (B). 10-5/16x5-3/16x6¾ in. high. (12 lbs.) . . . . 22.

D FM/AM MAHOGANY AIRCASTLE
• Incomparable FM and AM reception   **44.95**
• High fidelity PM dynamic speaker   or $5 month
• Beautiful rubbed mahogany cabinet
WONDERFULLY LIFE-LIKE—that's FM (Frequency Modulatior Makes it sound as though music, sports events or plays are ta ing place in your own home. Brings you round, full-bodi tone that you can't possibly receive on anything but an FM s Reception is startlingly interference-free even when atmc pheric conditions are unfavorable. Time-tested AM (Standa Broadcast) is better than ever. Brings in your favorite progra with wonderful clarity.
TOP-NOTCH RECEPTION ASSURED with sensitive supe heterodyne circuit. Power-packed 8-tube performance from preferred-type tubes (1 dual purpose) plus rectifier. Sensiti PM dynamic speaker equipped with Alnico No. 5 magnet—2 times more powerful than older types. Faithfully reproduces ton
BUILT-IN FM/AM AERIAL SYSTEM eliminates outside wirin Automatic volume control. Lovely mahogany veneer cabine Illuminated slide-rule dial for easier, sharper tuning. AM Ban 550-1600 kilocycles. FM Band: 88-108 megacycles. 105-125 Volt 60 cycles, AC only. You pay freight or express from Chicago
B76R 357. (D) FM/AM. 12x8x7¼ in. high. (14 lbs.) 44.9

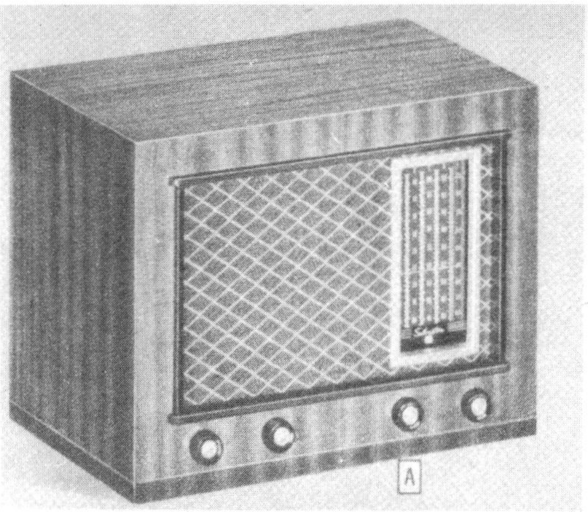

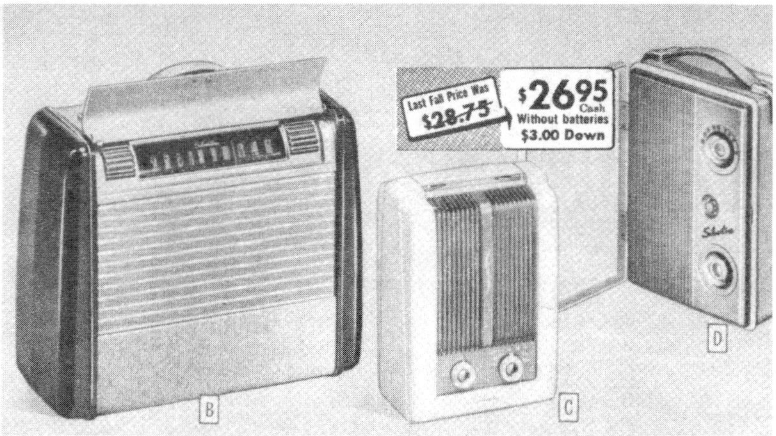

Last Fall Price Was $28.75
$26.95 Cash
Without batteries
$3.00 Down

NEW SILVERTONE with the greatest "reaching-power" ever!

[A] Now you can have a handsome Radio with more "reaching-power" on broadcast band and short wave than any other SILVERTONE ever! Tunes 540-1600 KC *plus* 4 excitement-packed short wave bands from 1.62-22.2 megacycles. Especially designed for those areas where normally only a few stations are picked up. "Climatized" to stand up anywhere from the tropics to the arctic. Full-toned 8-inch speaker; variable tone control. Input jack for record player. Outside aerial needed—use (E) or (F) on page 638.

$94.50 Cash
$8.00 Monthly

It's magnificent furniture, too! Hand-rubbed mahogany veneer cabinet. Size 17¼x10¾x13 in. For 110-125-145-200 or 245 volts, 40-60 cycle AC.
57 E 09054—UL approved. Shpg. wt. 30 lbs. $94.50

SILVERTONE PORTABLE RADIOS . . take entertainment and music everywhere you go . . they operate on AC or DC or on batteries

[B] **De luxe Portable.** Three tuning circuits (instead of the usual two) give you extra "reaching-power" in autos, trains, airplanes, steel buildings. Has 5 tubes (2-dual-purpose) plus rectifier; 5¼-inch speaker. Tunes 540-1600 KC.

Last Fall Price Was ~~$49.95~~
Less battery
$44.95 Cash
$5.00 Down

Sturdy aluminum case is specially painted to resemble rich leather. Shock-resisting brown plastic ends; built-in loop aerial. Plays with snap-down cover closed. Size 14x11x5⅝ in. Operates on 105-125 volts AC or DC or on battery pack (guaranteed for 175 service-hours). UL approved.
57 E 08270—Without battery. Wt. 14 lbs. $44.95
57 ET 8271—With battery. Shpg. wt. 21 lbs. . . 48.95

[C] **Personal Portable Radio.** Size 9¼x6¾x4 in. Ivory-color plastic case, gold-color grille; 4 tubes (2 dual-purpose) plus rectifier; built-in aerial; 3½-inch speaker. Operates on 117 volts AC or DC or on battery. UL approved.

WAS ~~$24.95~~
$23.95 Cash
Without battery
$2.50 Down

57 E 09260—Without battery. Shpg. wt. 6 lbs. $23.95
57 ET 9261—With battery. Shpg. wt. 7 lbs. $26.34

[D] **De luxe Personal Portabl Radio.** Size only 8⅜x6⅝x 3½ in. Built-in aerial; 4 tube plus rectifier; 3½-in. speaker Metal case has ivory-colo hinged plastic covers, attrac tive gold-color trim. Detach able leather shoulder strap Operates on 105-125 volts AC o DC or on batteries. For you protection, approved by Un derwriters' Labs., Inc.
57 E 08260—Without batteries Shpg. wt. 7 lbs. $26.9
57 ET 8261—With batteries. Shpg. wt. 8 lbs. $29.2

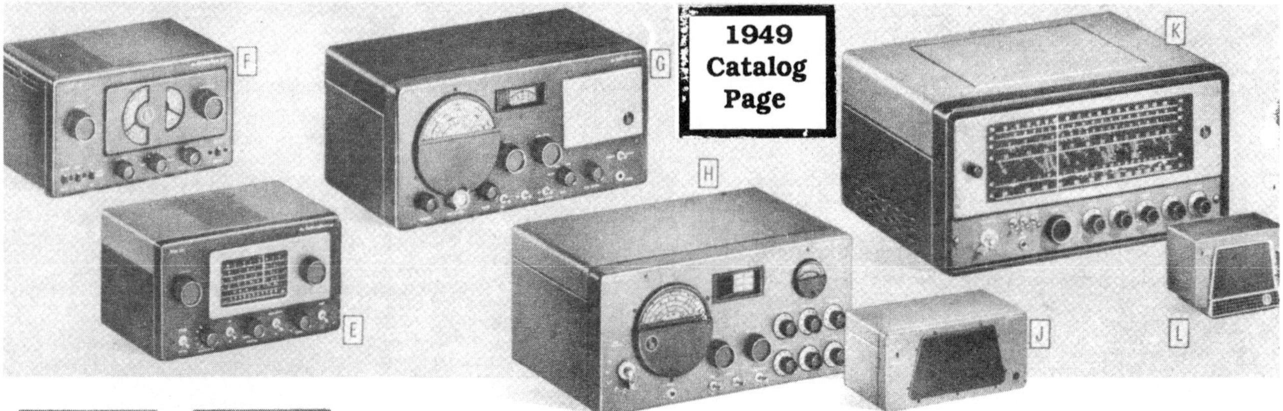

1949
Catalog
Page

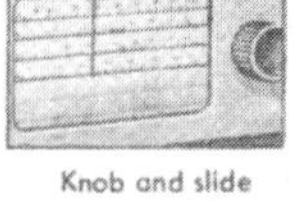
Knob and slide rule dial for regular tuning

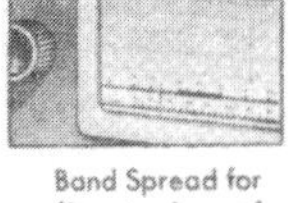
Band Spread for fine tuning of weak stations

HALLICRAFTERS . . the radio that amazes even the experts . . all the world's your neighbor—bringing entertainment to your own home!

[E] **Explore the fascinating invisible world of radio.** This Hallicrafters S-53 Receiver is crammed with a "reaching-power" and range that will amaze you. Tunes 540-1600 KC broadcast band *plus* 4 short wave bands from 1.6-54.5 megacycles. Two tuning controls . . . fine-tuning (Band-Spread) gives you the *exact* tuning you need for distant stations.

$89.50 Cash
$7.00 Monthly

Uses 7 tubes plus rectifier. Built-in 5-in. speaker with on-off switch. Volume and tone controls; Noise Limiter switch; Phone-CW switch for receiving code; jacks for headphones and for record player such as 57E08140 on page 630. Two-tone gray steel cabinet. Size 12⅞x7⅞x6⅞ in. For 105-125 volt, 50-60 cycle AC. Approved by Underwriters' Laboratories, Inc.
57 E 07451—Shpg. wt. 23 lbs. $89.50

[F] **Hallicrafters S-38 Receiver**... economy priced! Tunes 540-1600 KC *plus* 3 bands of shortwave from 1.6-32 megacycles. Has regular tuning plus fine tuning (Band Spread).

$49.95 Cash
$5.00 Monthly

Uses 5 tubes plus rectifier. Built-in 5-inch speaker with on-off switch. Volume control; Noise Limiter switch; Phone-CW switch for receiving code. Has jack for connecting headphones. Gray steel cabinet, 12⅞x7⅜x8⅝ in. Operates on 105-125 volts AC or DC. For your protection, approved by Underwriters' Labs., Inc.
57 E 07450—Shpg. wt. 14 lbs. $49.95

[G] **Hallicrafters S-40A Receiver.** Tunes 540-1600 KC *plus* 3 bands of short wave from 1.6-43 megacycles. Band Spread control for fine tuning. Uses 8 tubes plus rectifier. Built-in 5-in. speaker. Gray steel cabinet, 18½x9x11 in. For 105-125 volt, 60 cycle AC. UL approved.

$110.00 Cash
$9.00 Monthly

57 E 07456—Shpg. wt. 33 lbs. $110.00

[H] **Hallicrafters SX-43 Receiver.** Tunes 540-1600 KC *plus* 4 short wave bands from 1.6-55 megacycles *plus* 88-108 megacycle FM band. Gray steel cabinet, 18½x8⅞x13 in. Ten tubes plus rectifier. Separate Band Spread control for fine tuning. Operates on 105-125 volt, 50-60 cycle AC. UL approved. Speaker not included.

$189.50 Cash
$12.00 Monthly

57 E 07452—Shpg. wt. 45 lbs. $189.50

[L] 57 E 07453—6x9-inch oval speaker for above. Matching gray steel cabinet, 18½x8½x9⅞ in. Shpg. wt. 19 lbs. $24.50

[K] **Hallicrafters SX-62 de luxe Receiver.** Gives complete coverage of standard broadcast, FM and short wave in 6 bands from 540 KC to 108 megacycles. Has 14 tubes plus rectifier and voltage regulator. Gives brilliant, high-fidelity reception. Handsome slide rule dial . . . each band separately illuminated. Has regular tuning plus fine tuning (Band Spread). Gray steel cabinet, 20x10¼x18 in. Operates on 105-125 volt, 50-60 cycle AC. Approved by Underwriters' Labs., Inc. Speaker not included.

$289.50 Cash
$18.00 Monthly

57 E 07454—Shpg. wt. 65 lbs. $289.50

[J] 57 E 07455—8-inch speaker for above. Matching gray steel bass reflex cabinet, 12½x11¾x17 in. Shpg. wt. 30 lbs. $39.50

Shop the easy Mail Order way. For extra convenience, use Sears Easy Terms . . . see inside back cover

Leading the Portable Parade
Melrose
3 WAY PORTABLES

You've got the whole world of radio on the end of the smart handle of this versatile portable. Just seeing it makes people want it. Hearing it play is a thrill - it's packed with surprising power and rich with bell-like tone.

Plays Anywhere. This big hit in little portables carries and plays anywhere - indoors or outdoors - from 110 volts AC, DC or from self-contained batteries. Has built-in loop antenna for strong signal pickup. Tunes 535 to 1620 kc for full coverage of the standard broadcast band. The large slide-rule dial is clearly calibrated and extremely easy to tune.

Advanced Superhet Circuit. Precision-engineered to develop extra power and efficiency. Employs 4 latest type tubes plus rectifier. Has automatic volume control to maintain uniform volume level on all programs.

Superb Tone Quality. Just listen and judge - you'll be amazed at the exceptional clarity and fidelity of this compact performer. Incorporates the latest type PM Dynamic Speaker with powerful new Alnico V magnet. Has exceptional volume-carrying capacity and faithful reproducing quality.

Handsome Styling. Here's an irresistible package, styled in a smart, two-toned color combination. Strikingly designed in plastic, featuring streamlined surfaces and attractive curved speaker grille. Has practical, inset tuning knobs and sturdy carrying handle. Size: only 9½″ wide, 7¼″ high, 5⅝″ deep. Ship. wt., 5 lbs. (less batteries). Operates on 110 volts AC or DC, or batteries. Underwriters' approved.

No. 2804. List $31.00 less batteries. Dlr's., ea. $19.81. 3 lots, ea. $18.58, less 2%, net...................... **18²¹**

Batteries For Above Portable

No. 2791. 67½ volt "B" (1 required). List $2.45. Dlr's., ea. $1.88. less 2%, net...**1.84**

No. 2801. 4½ volt "A" (1 required). List 75c. Dlr's., ea. 57c. less 2%, net **56c**

Works on
110-volt AC
110-volt DC
Batteries

1949
Catalog
Page

...tainment
...t-of-doors

Wake up to Music with the....

Melrose
WAKEMASTER
The Combination
Radio-Alarm CLOCK

5 Tubes Including Rectifier
Built-in Air-O-Scope Antenna
4-In. Alnico V PM Speaker
Automatic Volume Control
Beautifully Styled Plastic
Cabinet

Equipped with
Telechron
ALARM MOVEMENT

Everybody wants this exciting combination - a famous Telechron clock and a marvelous, pure tone radio in one streamlined cabinet. Just dial a station, set the alarm, and the radio automatically goes on at the pre-set time. A completely independent alarm in the clock rings ten minutes later in the event sleeper has not been wakened by radio. Size 10⅝″ long, 6¼″ high. 105-125 volts, 60 cycle, AC. Wt. 5½ lbs.

Walnut Cabinet	Ivory Cabinet
No. 2619. List $31.95. Dlr's. $23.10. 3, ea. $21.73, less 2%........ **21.30**	**No. 2602.** List $33.95. Dlr's. $24.25. 3, ea. $22.75, less 2%........ **22.29**

Melrose FM-AM RADIO

True Console Type FM Reception
8 Tubes Including Rectifier
Will Receive in Hard-to-get Areas
Illuminated Slide Rule Dial
Loop Antenna for AM Reception
Line Antenna for FM Reception

Here is static-free, console tone FM reception plus excellent, high-fidelity standard reception . . . both contained neatly and compactly in a streamlined plastic cabinet. The tone control adjusts to razor-fine fidelity. The large slide rule illuminated Vernier dial is easy to see . . . easy to tune. Large PM dynamic speaker produces clearest quality of sound. 8-tube engineering effectively solves the problem of tough-reception areas. Needs no aerial. With all the big-radio listening advantages and the added convenience of space-saving size, this value-packed set has outstanding sales appeal. 11½″ long, 7½″ high. AC or DC. Wt. 8½ lbs. **32⁷⁸**

No. 2747. Walnut. List $49.95. Dlr's., ea. $36.70. 3 lots, ea. $33.45, less 2%, net...................................

Sonora 6 Volt Battery Sets

Clear as a Bell

Sonora-Model M-22
4 TUBE-6 VOLT SUPERHET

- 8 Tube Performance
- Beam Power Output
- Illuminated Slide Rule Dial
- Synchronous Vibrator
- Automatic Volume Control
- Low Battery Drain
- No "B" or "C" Batteries
- Striking Bakelite Cabinet

$15 28 Net

In Lots of 3

An exceptional performer, equal in every detail to ordinary 8 tube receivers. Employs a superheterodyne circuit of advanced design unusually sensitive and selective. The use of dual and triple purpose tubes, plus the dual purpose synchronous vibrator, gives this set the power and range of the largest battery receivers. Tunes from 535 to 1720 K.C.'s (includes 1712 K.C. police channel) on a full vision illuminated slide rule dial. Requires only one 6 volt storage "A" battery. No "B" or "C" batteries. Unusually low drain of 1.9 amps.—actually 50% less than that consumed by other battery receivers of equal operating efficiency. Special permanent magnet dynamic speaker requires no battery current. Complete with the following tubes: 1—6A8G, 1—6K7, 1—6Q7G, 1—6G6G. Gorgeous two-tone bakelite cabinet in fluted design with special honey-comb-effect grille. Front in ___ in black. 11¾" long, 7½" deep, 7¼" hi___
No. 9H4286. List $24.95. Dlr's., e___
$16.11. Lots of 3. ea. $15.59, less 2%, ___

6 Station Automatic Push Button Tuning

Sonora-Model QA-33
6 TUBE-6 VOLT SUPERHET

- No "B" or "C" Batteries Required
- Gets Foreign Stations
- 6 In. Slide Rule Dial
- Low Battery Drain
- Automatic Volume Control
- Variable Tone Control
- 9-Tube Performance

Here is a set that offers your battery radio prospects all the thrilling conveniences usually found only in electric sets. 6 Station Automatic tuning—Foreign Reception—to mention but a few of the features that make modern radio a joy to own. Dual wave bands —from 535 to 1720 K.C.'s, and from 5650 to 18,000 K.C.'s bringing in Foreign and Domestic Short Wave, as well as the regular broadcasts. **Extremely low battery drain,** of only 1.85 amps. from only one 6-volt storage battery—no "B" or "C" batteries required. 6" Permanent Magnet Dynamic Speaker. Multiple purpose tubes and a dual purpose synchronous vibrator result in 9-tube performance. 2 watt output. **Full A. V. C.** and continuously variable tone control. Complete with the following tubes: 1—6C8G, 1—6H4G, 1—6F7G, 1—6G7G, 1—6L5G, 1—1J6G. Smartly styled cabinet in latest 1939 design with attractive corner type grill. Horizontal overlay trim of vertically grained walnut striping crosses the front face. Instrument panel is of quarter-cut choicely grained walnut. 17½" long, 8" deep, 7¼" high. Wt. 22 lbs.
No. 9H4105. List $49.95. Dlr's., ea. $28.00. Lots of 3, ea. $26.65, less 2%, net **26 12**

3 BANDS- *World-Wide Reception!*
Sonora-Model 170-6B
6 TUBE-6 VOLT SUPERHET

An unusual value in a precision built 6-volt Receiver. 3 full bands provide a world wide range of reception. Superbly designed Superheterodyne circuit. Employs a synchronous, self-rectifying vibrator which combines the functions of an alternator and rectifier, thereby obtaining the equivalent of 7-tube performance. Equipped with full automatic volume control and continuously variable tone control. Undistorted output of 2.1 watts. Low battery drain of 1.8 amps. Full vision airplane type dial. Complete with R.C.A. tubes as follows: 1—6D8G, 1—6S7G, 1—6T7G, 2—30, 1—19. Handsome laydown type cabinet with top, front and bottom made from single piece of richly grained walnut. Rolled effect top and bottom. Hand rubbed and highly polished. Dimensions: 19" long, 11½" wide, 10" deep. Wt. 24 lbs.
No. 9H4243. List $49.95. Dlr's., ea. $22.50. Lots of 3, ea. $21.25, less 2%, net **20 82**

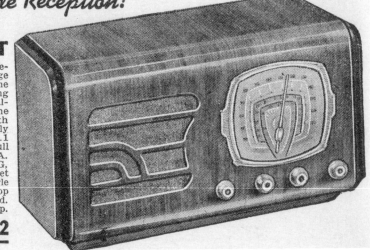

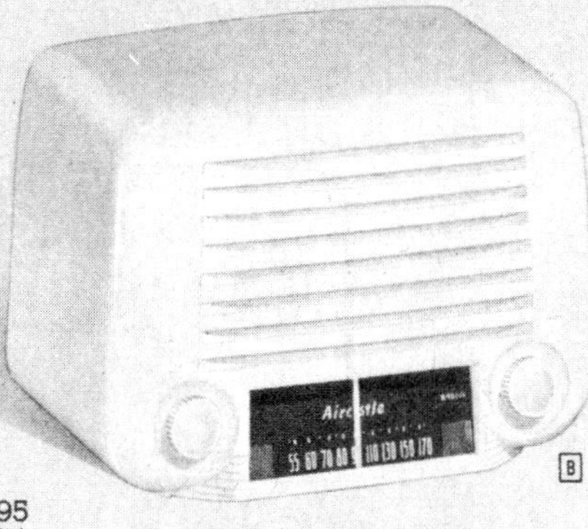

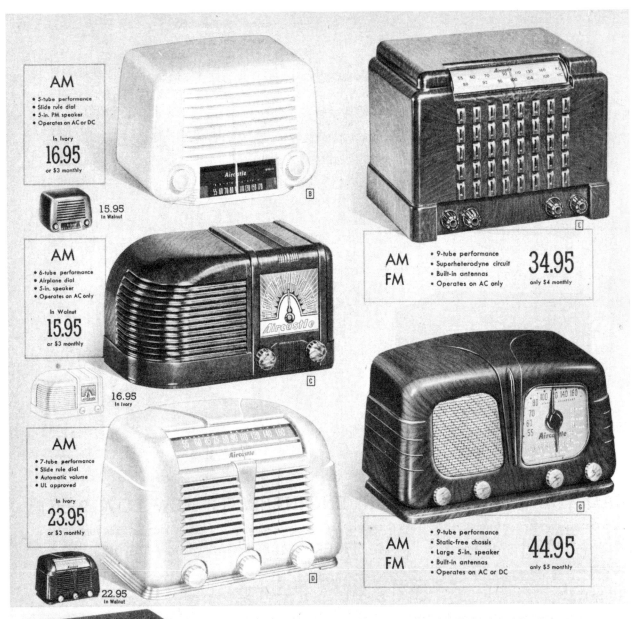

AM

- 5-tube performance
- Slide rule dial
- 5-in. PM speaker
- Operates on AC or DC

In Ivory
16.95
or $3 monthly

[B]

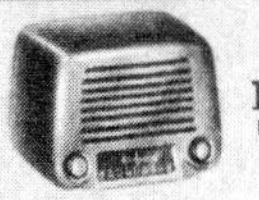

15.95
In Walnut

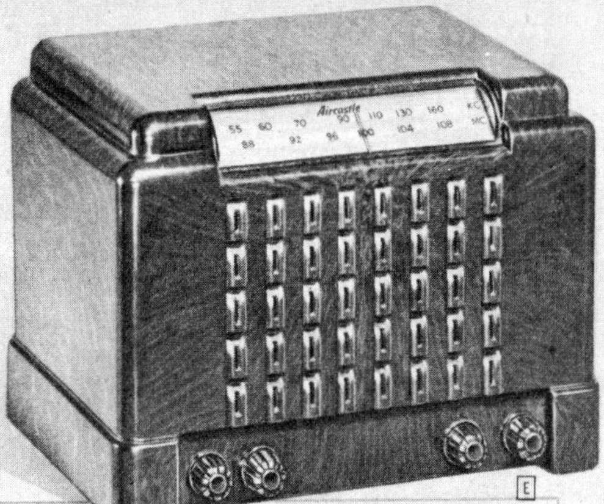

[E]

AM FM

- 9-tube performance
- Superheterodyne circuit
- Built-in antennas
- Operates on AC only

34.95
only $4 monthly

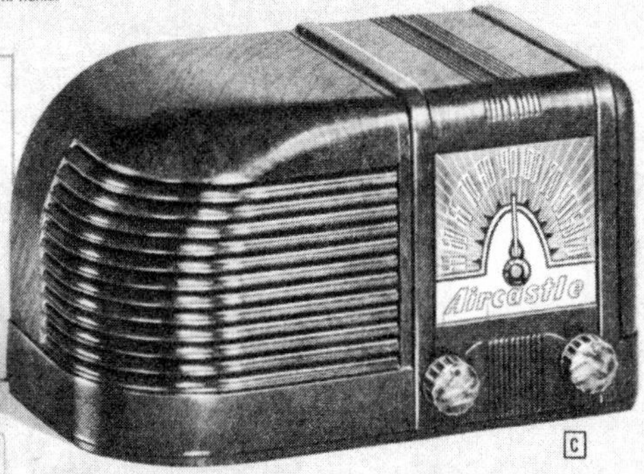

AM

- 6-tube performance
- Airplane dial
- 5-in. speaker
- Operates on AC only

In Walnut
15.95
or $3 monthly

[C]

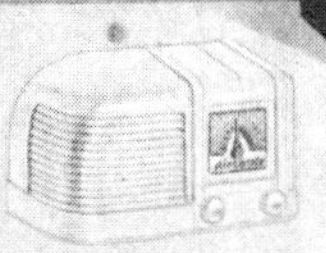

16.95
In Ivory

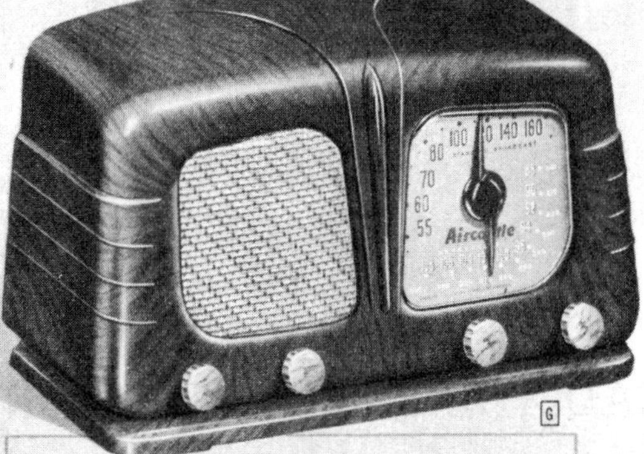

[G]

AM FM

- 9-tube performance
- Static-free chassis
- Large 5-in. speaker
- Built-in antennas
- Operates on AC or DC

44.95
only $5 monthly

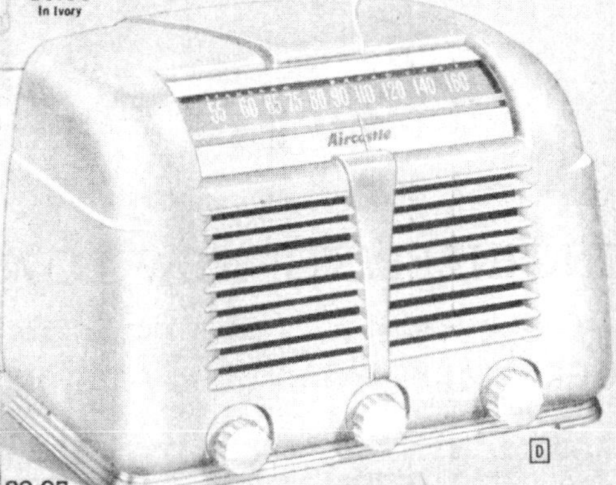

AM

- 7-tube performance
- Slide rule dial
- Automatic volume
- UL approved

In Ivory
23.95
or $3 monthly

[D]

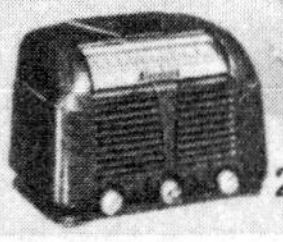

22.95
In Walnut

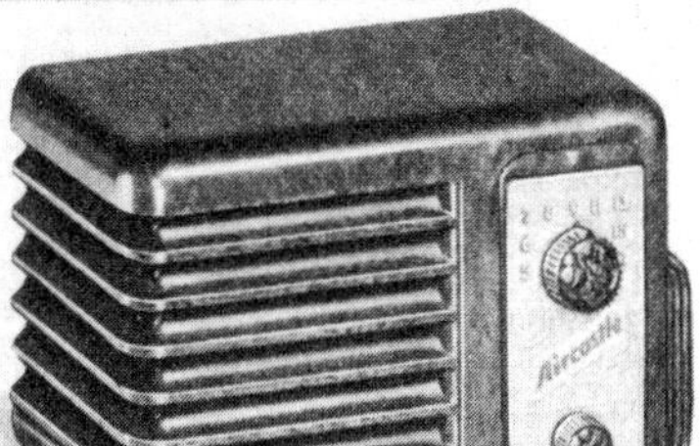

9.69 [A]
In walnut

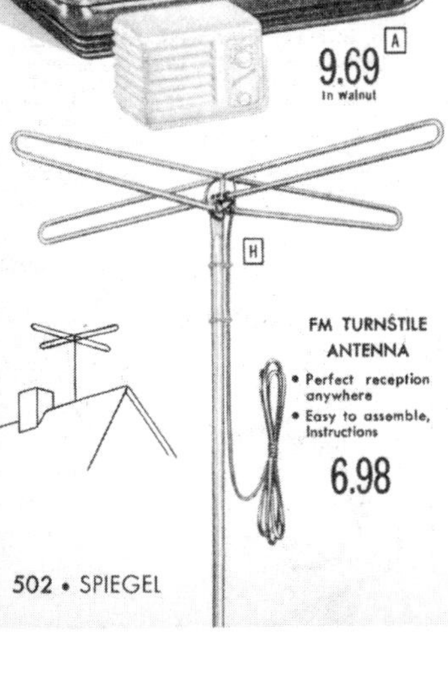

[H]

FM TURNSTILE ANTENNA

- Perfect reception anywhere
- Easy to assemble, Instructions

6.98

AIRCASTLE TABLE RADIOS PRICED LOW

[A] BUDGET MIDGET AIRCASTLE

- Extra sensitive superheterodyne circuit
- 4-in. PM dynamic speaker—Alnico magnet
- Easy-to-read dial—bright plastic cabinet

USE IT ANYWHERE IN YOUR HOME. Packs 4-tube performance into 3 tubes (1 dual purpose tube) plus rectifier. Direct drive tuning. Hank antenna. Size: 7¼x4½x5¼ in. Standard Broadcast Range: 540-1720 KC. 105-120 V., AC or DC. (Not recommended 50 miles from broadcasting station.)
A76S 374. (A) In Walnut. (7 lbs.)...... **9.69**
A76S 375. In Ivory (7 lbs.)............10.69

[H] FM ANTENNA—TURNSTILE TYPE

Perfect reception, from all area stations. With 5-ft. weatherproof mast, universal adjustable mounting base, 60-ft. lead wire, 4 stand-offs for side of building, 6 plastic insulators. Easy to assemble. Instructions.
A76S 1389. (H) Shpg. wt. 9 lbs. Mailable. **6.98**

[B] AIRCASTLE TABLE MODEL

- 5-tube performance from 4 tubes, rectifier
- Extra powerful superheterodyne circuit
- PM dynamic speaker—built-in loop antenna

COMBINES SUPERB RECEPTION AND BEAUTY into a thrifty Aircastle—guaranteed to please. Comparison proves—dollar for dollar . . . value for value . . . Aircastle is tops for all-around radio enjoyment.

SENSITIVE SUPERHETERODYNE CIRCUIT incorporates 5-tube performance from 4 tubes (1 dual purpose tube) plus rectifier.

LARGE 5-IN. PM SPEAKER—with the heavy Alnico No. 5 magnet—2½ times more power-ful than others. Insures better rec reproduction. Takes less power 7½-to-1 vernier tuning. 8½x6⅛x5 Standard Broadcast range: 540 t 105-120 volts, 60 cycles; AC or D weight 7 lbs.
A76S 378. (B) Walnut plastic cabi
A76S 379. Ivory plastic cabinet...

[C] HIGH QUALITY AIRCASTLE

- Richer toned—big 5-inch dynamic speaker
- Built-in loop antenna—external connection
- Lighted dial face—8-to-1 vernier tuning

INSURED HOURS OF LISTENING PLEASURE—packed into one compact Aircastle table model radio. Hear it once and you'll agree—there is no greater value in radios today!

POWERFUL 6-TUBE PERFORMANCE from 4 advanced type tubes (2 dual purpose) plus rectifier. Big 5-in. PM speaker reproduces tone with superb fidelity. Equipped with the famous Alnico No. 5 magnet.

INTERFERENCE-FREE RECEPTION from the sensitive superheterodyne circuit. Built-in loop antenna . . . external antenna connection for use in out of way places. Easy-read illuminated dial. 12x6¼x7 in. high. Standard Broadcast range: 540-1700 KC. 105-120 volts, 60 cycle, AC only. Shipping weight 9 lbs.
A76S 380. (C) Walnut plastic cabinet. **15.95**
A76S 381. Ivory plastic cabinet........16.95

[D] NEW AIRCASTLE TABLE RADIO

- More sensitivity of superheterodyne circuit
- Volume control stops fading and blasting
- Approved by the Underwriters' Laboratories

AIRCASTLE PRESENTS its new and powerful table model radio. Compactly built to occupy little space . . . yet give the overall performance that has made Aircastle America's finest radio in this price range today!

POWERFUL 5-IN. PM DYNAMIC SPEAKER—equipped with the heavy Alnico No. 5 magnet! Automatic volume control eliminates fading or blasting of far-away stations. Incorporates 7-tube performance from 5 tubes rpose tubes) plus rectifier. Built-enna.

ERODYNE CIRCUIT for greater and selectivity. Slide rule dial. ¼ in. Broadcast range: 535 to 1620 V., AC or DC. (14 lbs.)
D) Walnut plastic cabinet. **22.95**
vory plastic cabinet........23.95

[E] AM/FM AIRCASTLE RADIO

- Variable tone control for true reproductio
- 9-tube performance; superheterodyne circu
- Easy-to-read illuminated slide rule dia

DOUBLE YOUR LISTENING ENJOYMEN with FM (Frequency Modulation) . . . designe to give static-free reception even under un favorable atmospheric conditions. Dual con trols. Tune in on the new FM programs or ol favorites on AM (Standard Broadcast).

9-TUBE PERFORMANCE from 7 tubes (2 du purpose) plus rectifier. Superheterodyne cir cuit insures greater sensitivity and selectivity Built-in loop antenna—external connection for difficult reception areas.

5-IN. PM SPEAKER—features heavy Alnic No. 5 magnet. Walnut plastic cabinet. 15¼ 8½x9 in. high. Standard Broadcast: 540-162 KC. FM: 88-108 megacycles. 105-120 volt 60 cycles, AC only. You pay frt. or exp. fro Chicago. Shipping weight 15 lbs.
B76S 382. (E) AM/FM Radio. **34.9**

[G] DELUXE AM/FM AIRCASTLE

- 5-in. PM dynamic speaker—Alnico magne
- Automatic volume, variable tone control
- Greater sensitivity; superheterodyne circui

PERFECTION IN HOME ENTERTAINMENT Combines the crystal clear, static-free recep tion of FM (Frequency Modulation) with time tested AM (Standard Broadcast)!

SUPERHETERODYNE CIRCUIT insures greate sensitivity and selectivity with fewer tubes Powerful performance of 9 tubes from 7 tube (2 dual purpose) plus rectifier. 9½-to-1 vernie tuning ratio—illuminated airplane dial.

HEAVY ALNICO No. 5 MAGNET incor porated into large 5-in. PM dynamic speaker—2½ times more powerful than other types Walnut finish plastic case. 8½x14x8 in. high Standard Broadcast: 540-1620 kilocycles. FM 88-108 megacycles. 105 to 120 volts, AC or DC You pay freight or express from Chicago Shipping weight 17 lbs.
B76S 383. (G) Deluxe AM/FM Radio. **44.9**

1950 Catalog Page

Low Priced Silvertone *Portable Radios*

. . . take along a bit of spring wherever you go

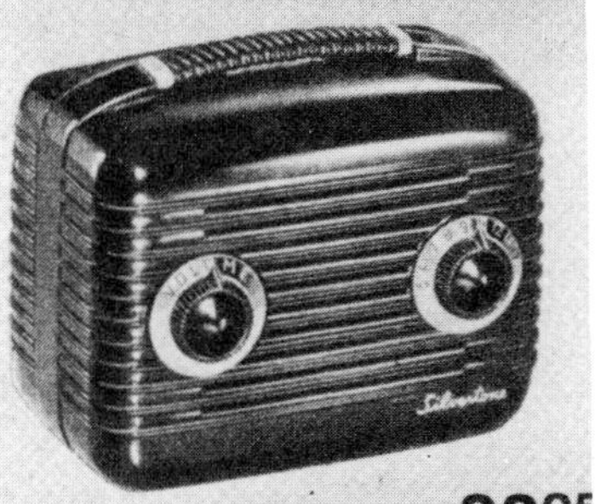

Without battery $22^{95}

3-WAY PLAYMODEL PORTABLE RADIO

- Plug in . . use as smart table radio
- Lightweight . . easy to carry anywhere
- Battery easy to replace . . quickly

You'll find it hard to believe that you can own this brilliant performing 3-way portable *at such a down-to-earth thrifty price!* Take it along everywhere you go . . . use it 3-ways—on AC, DC or battery.

Simple direct tuning. Properly baffled 3½-in. speaker. 4 tubes plus rectifier. Built-in antenna. Tunes 540-1600 KC. Beautiful maroon plastic cabinet, 8¾x7x4⅞ in. For 105-125-volt, 50-60 cycle AC-DC. Approved by Underwriters' Laboratories, Inc.

57K0215—Without battery. Shipping weight 5 pounds. $22.95
57 KT 216—With battery. Wt. 6 lbs. 24.95

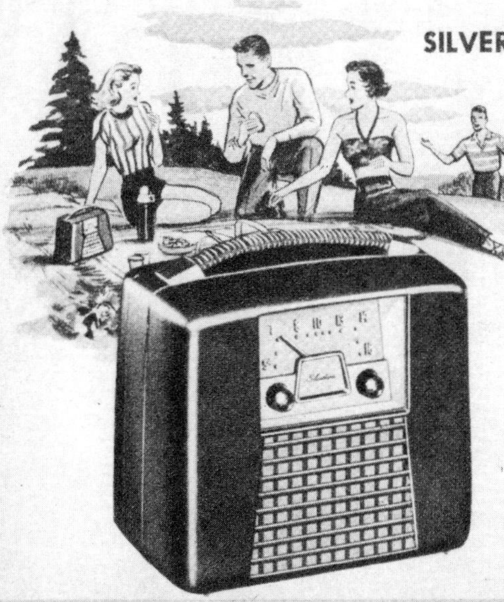

SILVERTONE 3-WAY SUPER PORTABLE

29^{95} Cash — Without battery — $3.00 Down

- Easy sweep, indirect tuning dial
- Big, full-voiced 6x4-in. speaker for richer, more thrilling tonal quality
- Uses longer-life 150-hr. battery

There's always a breath of spring in the air when you take along this compact, sleek, lightweight SILVERTONE Portable Radio . . . take it to picnics, gay springtime parties, sporting events, friends' homes, whisk it away to any room! And you can use it 3 ways, too . . . on AC, DC or as a battery portable. Large, powerful 6x4-inch oval speaker. Has 4 SILVERTONE tubes plus rectifier. Exclusive SILVERTONE Radionet built-in antenna . . . super-sensitive to pull in more station signals. Tunes 540-1600 KC standard broadcast. Maroon plastic cabinet, 10¾x9½x6⅛ in. For 105-125-volt, 50-60 cycle AC-DC. UL approved.

57 K 0220—Without battery. Shpg. wt. 8 lbs. $29.95
57 KT 221—With battery. Shpg. wt. 13 lbs. 34.40

OUR LOWEST PRICED PORTABLE RADIO

Without battery $14^{95}

Instant playing SILVERTONE battery-operated portable. Handsome marbleized green plastic cabinet, 6½x8⅞x3½ inches. Controls recessed in each end of cabinet . . . easy direct tuning. 4 tubes; 4-in. speaker. Tunes 540-1600 KC standard broadcast. Built-in antenna.

57K0210—Without battery. Wt. 4 lbs. $14.95
57KT211—With battery. Wt. 5 lbs. 16.95

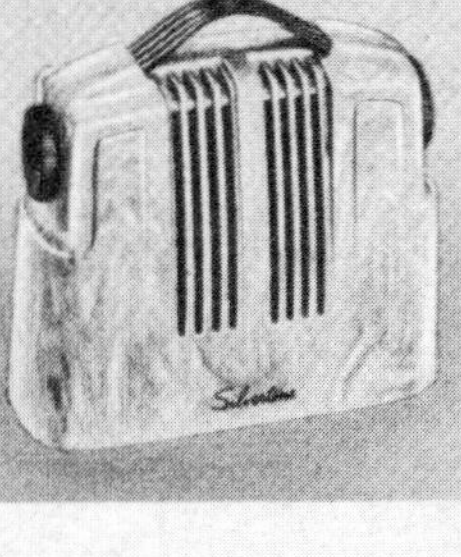

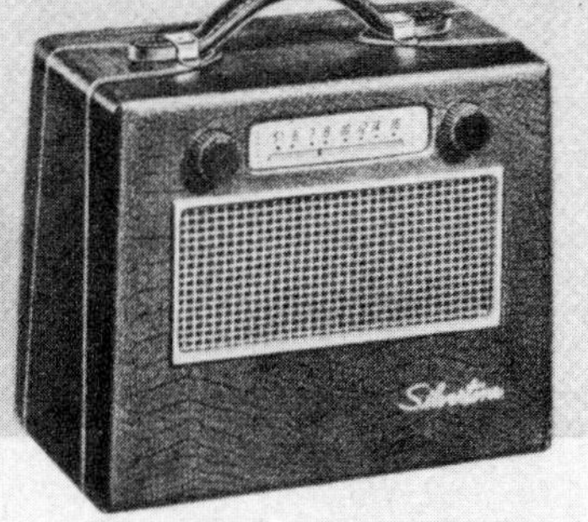

SILVERTONE'S FINEST 3-WAY PORTABLE RADIO

Without Battery 36^{95} Cash — $4 Down

Our most powerful portable for best reception. Uses long-life 175-hr. battery or 105-125-volt, 50-60-cycle AC or DC. *Triple-tuned circuits* for greatest sensitivity. 5 tubes, rectifier. *5¼-in. speaker.* Built-in antenna. 540-1600 KC. Tan artificial leather cover. 11½x9⅛x6 in. UL approved.

57K0225—Without battery. Wt. 10 lbs. . . . $36.95
57KT226—With battery. Wt. 17 lbs. 41.60

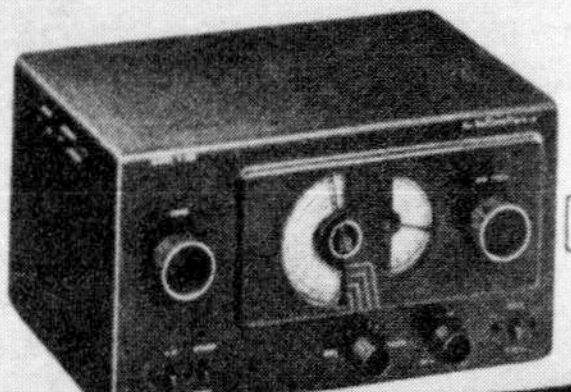

A

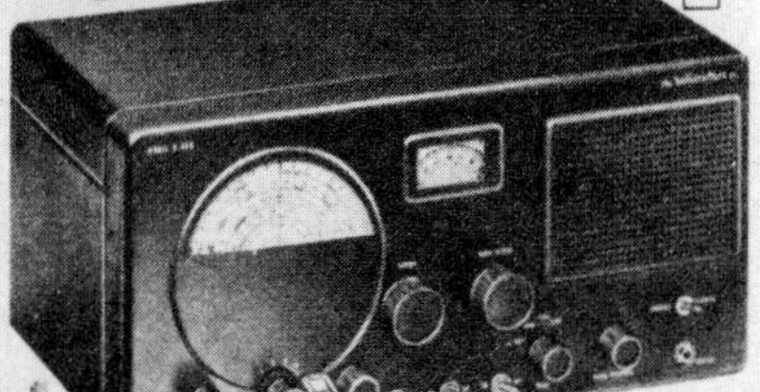

B

HALLICRAFTERS . . world famous for powerful reception

S-38B Receiver

49^{50} Cash — [redacted] — $5 Monthly

[A] Highly sensitive on 540-1600 KC standard broadcast plus 3 bands of short wave from 1.6 to 32 megacycles. 1-watt power output at less than 10% distortion. Regular plus fine tuning. 4 tubes plus rectifier. Built-in 5-in. speaker with on-off switch. Phone-CW switch for receiving code. Headphone jack. Dark gray steel cabinet, 12⅞x7x . . . -125-volt, 50-60- . . . C. UL approved. . . . n, we recommend . . . system 57K6705, . . . pping weight 13 . . . . $49.50

1951 Catalog Page

S-40B Receiver

99^{95} Cash — [redacted]

[B] *A more powerful, more sensitive* Hallicrafters to let you enjoy good radio reception . . . regardless of where you live! 2-watt power output at less than 10% distortion. 540-1600 KC standard broadcast . . . plus 3 short wave bands from 1.6 to 44.0 megacycles. 2 tuning controls . . . fine tuning for hairline adjustment. Has 7 tubes plus rectifier. Built-in 5-in. speaker with on-off switch. Phone-CW switch; headphone jack. Gray steel cabinet, 18½x9x9 in. 105-125-volt, 50-60-cycle AC. UL approved. *Order Deluxe Antenna System listed below.*

57K07456—Shpg. wt. 32 lbs. $99.95

S-72 and S-72L Portables

109^{95} Cash — [redacted] — Model S-72

[C] 3-way operation . . . AC, DC or battery power. One RF, two IF stages. 540-1600 KC standard broadcast plus 3 short wave bands from 1.6 to 30.0 megacycles. Bandspread for precision tuning. 8 tubes plus rectifier. Built-in 5-in. speaker. Loop antenna plus telescoping antenna for short wave. 105-125-volt, 50-60-cycle AC-DC. Tan imitation leather covered wood case, 14x12¼x7¼ in. UL approved.

57K07457—Hallicrafters S-72. Shipping weight 16 pounds. $109.95
57K07459—Hallicrafters S-72L. Same as above except *covers Weather Band . . . 180-400 KC*; standard broadcast, 540-1600 KC; 2 short wave bands from 1.6 to 11.5 MC. Shpg. wt. 16 lbs. . . $119.95

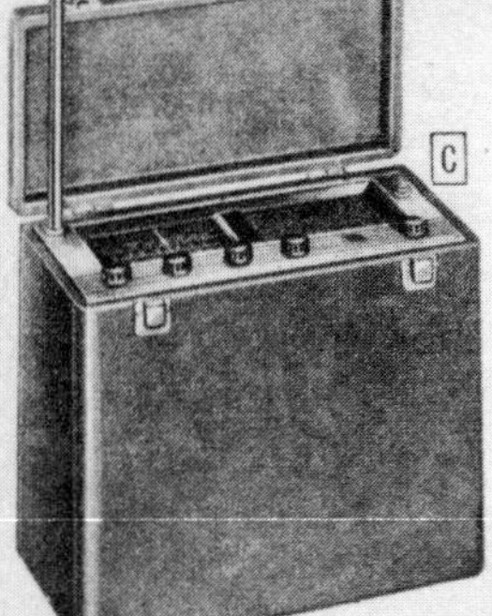

C

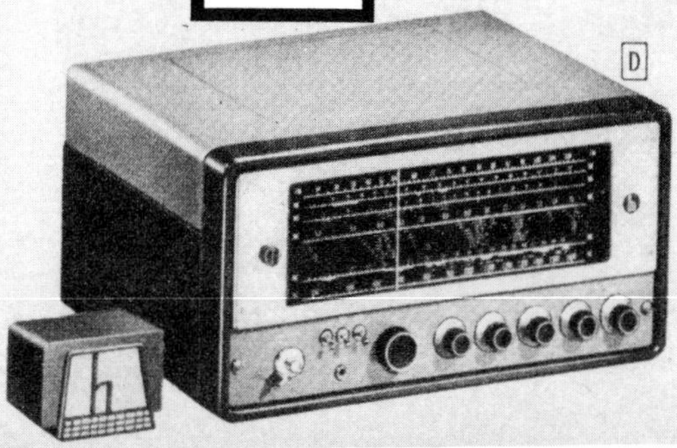

D

SX-62 Receiver . . our most powerful!

289^{50} Cash — $75 Down — $17 Monthly

[D] *The finest Hallicrafters that we offer!* Has tremendous reaching power and range. Tunes 540-1600KC standard broadcast, *5 short wave bands* from 1620 KC to 109 MC PLUS *highly stable FM reception* from 27 to 108 MC! 8-watt power output at less than 8⅓% distortion. Has 14 tubes plus voltage regulator and rectifier. *Six degrees of selectivity.* Two RF and two IF stages for distance reception. Handsome gray metal cabinet, 20x10¼x16 in. For 105-125-volt, 50-60-cycle AC. Approved by Underwriters' Laboratories, Inc., for your safety and protection. *We recommend using antenna and speaker listed below.* Order the best . . . it's your best buy!

57 KM 7454—Shipping weight 66 pounds. $289.50
57 K 6705—Deluxe Antenna System, complete. Shpg. wt. 5 lbs. 5.49
57 K 07460—Speaker R-46 for SX-62. 10-in. Shpg. wt. 30 lbs. 19.95

SILVERTONE AT ITS BEST — OUR MOST POWERFUL AM-FM TABLE RADIO!

Picks up *more* stations on *both*

AM and FM bands and *holds* them

. . . tops in stability, selectivity

Exclusive "Roto-band" Tuning...Wide spread, sweep dial for greater accuracy

Iron Core Tuning ...cuts down annoying station drift

Ideal Fringe Area Reception . . . available with Outdoor Antenna

Brown, without antenna **$37⁹⁵** Cash
$4.00 Down

Ivory-color, without antenna **$39⁹⁵** Cash
$4.00 Down

Just a flick of the knob on this power-packed SILVERTONE and you have standard AM reception or the fascinating world of FM at your command!

Designed to pick up *more* stations, especially in remote fringe areas *and hold them!* Exclusive "Roto-band" *wide-spread, sweep-dial* gives faster, accurate tuning on AM (540-1600 KC) standard broadcast and FM (88-108 Megacycles). *Iron core tuning* cuts station drift and gives more selectivity—greater stability. Large 5-inch speaker. 7 tubes plus rectifier. Built-in loop AM—built-on FM antennas. *For best FM reception*, particularly in fringe areas, we recommend the radios with outdoor antennas. Beautiful plastic cabinet, 7⅜x12¾x7-inches. For 105-125-volt, 50-60-cycle AC only. UL approved. *Mailable.*

57 K 018—Brown cabinet. Shipping weight 11 lbs.............. $37.95
57 KT 19—Above with outdoor antenna. Shipping wt. 23 lbs........ 47.75
57 K 020—Ivory-color cabinet. Shipping weight 11 lbs............. 39.95
57 KT 21—Above with outdoor antenna. Shipping wt. 23 lbs........ 49.75

VERSATILE SILVERTONE CLOCK-RADIO

1951 Catalog Page

Automatically wakes you to music...the perfect way to start the day!

Gently lulls you to sleep . . . sleep-switch turns radio off!

Times small kitchen appliances...controls many household items!

In rich-looking brown plastic cabinet **$25⁹⁵** Cash $3.00 Down

In smartly styled ivory-color plastic **$27⁹⁵** Cash $3.00 Down

It's sensational! SILVERTONE's dual-purpose "Lullalarm" *gently* nudges you awake . . soothes you to sleep to the voice of a radio! You simply pre-set the alarm and the radio automatically wakes you at the appointed hour . . . *buzzer reawakens sleepyheads after 10 minutes!* It sings you to sleep . . . the sleep-switch turns radio off automatically!

Has continuously operating electric clock and rich-voiced AM radio. All clock controls are in cabinet front, set in dial face. Radio tunes 540-1600 K.C. 4 tubes plus rectifier. 4-in. dial-type speaker. Built-in antenna. Cabinet size, 10⅝x5¾x5¼ inches. For 105-125-volt, 60-cycle AC only. UL approved.

57 K 010—Brown plastic cabinet. Shipping weight 7 pounds.................$25.95
57 K 011—Ivory-color plastic cabinet. Shipping weight 7 pounds............ 27.95

LOW COST, COMPACT SILVERTONE

Large circular dial . . . directly tunes station in a wink!

Built-in loop antenna... does away with unsightly hanging aerial wires

Sturdy chassis, compact and well constructed

Attractive brown plastic cabinet **$15⁹⁵**

Fashionable ivory-color plastic cabinet **$16⁹⁵**

Unusually sensitive to station signals for so compact a radio! You can be justly proud of this glamorous-looking SILVERTONE in any room of your home. You'll like the smoothly molded plastic cabinet...the amazingly good reception and fine tone . . . the easy-to-tune large circular dial . . . *and* the low, low price that's so mighty budget-pleasing!

4-inch permanent magnet speaker gives full-toned volume. 4 SILVERTONE tubes plus rectifier. Built-in loop antenna. Tunes 540-1600 KC. Cabinet size, 8⅝x5¾x6 in. For 105-125-volt, 25-60 cycle AC or DC. Approved by Underwriters' Laboratories, Inc. for your protection. *Mailable.*

57 K 09000—Brown plastic cabinet. Shipping weight 5 pounds.............$15.95
57 K 09001—Ivory-color plastic cabinet. Shipping weight 5 pounds.......... 16.95

TRAVLER PLASTIC TABLE RADIOS

TRAV-LER 5-TUBE RADIO

Latest advanced type superheterodyne model. Has large speaker with extra powerful Alnico magnet for finest tone quality. Finest tone range from low bass to high treble. A two-tone lustrous plastic cabinet of outstanding beauty. The choice of contrasting and matching dial grille will please even the most critical. Large bright gold finish dial pointer and matching knobs add both a rich appearance and ease of tuning. Automatic volume control circuit eliminates blasting and fading. Has newest built-in type "Ferrite" iron rod antenna. 4 radio tubes and rectifier. Includes twin dual purpose tubes giving **six-tube performance.** Large semi-circular dial is easy to read and easy to tune. Specify color. Tuning range 540 to 1610 KC. 105-125 volts, 50 cycles, AC and DC.

No. 1RT-55-38M (Mahogany). ....(Index T1-1782, T3-1697).......List **22.95**
No. 1RT-55-38IR (Ivory and Red)............(Index T1-2098, T3-1997).......List **27.95**
No. 1RT-55-38R (Red). ....(Index T1-2098, T3-1997).......List **27.95**
No. 1RT-55-38IG (Ivory and Green)........(Index T1-2098, T3-1997).......List **27.95**
No. 1RT-55-38GY (Green and Yellow).....(Index T1-2098, T3-1997).......List **27.95**

TRAV-LER 5-TUBE ADVANCED MODEL PLASTIC TABLE RADIO

Wide selection in choice of color in this TRAV-LER PLASTIC TABLE MODEL RADIO permits matching modern designed set with any room setting. Balanced speaker and dial openings form a very attractive front. Large speaker with powerful Alnico magnet for finest tone quality. Finest tone range, bringing in clearly and distinctly the lowest bass or the highest treble. The set is the latest advanced type superheterodyne and is licensed under RCA and Hazeltine patents. Has four radio tubes and rectifier. Includes **two dual purpose tubes** giving six-tube performance. Large easy to read gold finish dial. Calibrated for easy tuning with civilian defense emergency frequency markings. Measures 9½" W x 6" H x 5" D. Operates on 105-125 volts, 50-60 cycles, AC or DC. Standard broadcast band from 540 to 1610 KC. Built-in sensitive "Ferrite" iron rod antenna.

No. 1RT-55-37M (Mahogany). ....(Index T1-1498, T3-1426).........List **17.95**
No. 1RT-55-37I (Ivory). ....(Index T1-1572, T3-1497).........List **19.95**
No. 1RT-55-37R (Red). ....(Index T1-1572, T3-1497).........List **19.95**
No. 1RT-55-37G (Green). ....(Index T1-1572, T3-1497).........List **19.95**

TRAV-LER 6-TUBE "JEWEL" PLASTIC TABLE RADIO

Modern cabinet design in rich jewel-like plastic. Pleasing design is available in assorted plain and mottled decorator colors. The handsome gold finish dial and knobs contrast pleasantly and effectively with dail pointer. Set is equipped with the latest in antennas, the "Ferrite" iron rod antenna. The large speaker has the extra powerful Alnico magnet for finest tone quality. No distortion on the highest treble or the lowest bass tone. Clear and bell-like reception. The large semi-circular dial is the center of interest in the overall cabinet design. The large dial numbers make tuning easy and pleasant. With 3-gang condenser. Tuning range: Standard broadcast band from 540 to 1610 KC. Automatic volume control eliminates blasting or fading. Operates on 105-125 volts, 50-60 cycles, AC or DC. Measures 12½" W x 7" H x 5¾" D.

No. 1RT-65-40M (Mahogany) ....(Index T1-2622, T3-2497)............List **34.95**
No. 1RT-65-40I (Ivory). ....(Index T1-2937, T3-2797)............List **39.95**
No. 1RT-65-40R (Red). ....(Index T1-2937, T3-2797)............List **39.95**
No. 1RT-65-40G (Green). ....(Index T1-2937, T3-2797)............List **39.95**
Nc. 1RT-65-40B (Blonde). ....(Index T1-2937, T3-2797)............List **39.95**

TRAV-LER 5-TUBE SATIN PLASTIC TABLE RADIO

Beautiful smooth-as-satin plastic with pleasing rounded corners and modern design front. Large size for tonal quality. Choice of rich colors in Red, Ivory, Green or Mahogany. Set is equipped with a large speaker assuring extra tone quality. Also has extra powerful Alnico magnet. The dial numbers are extra large, making tuning in programs easy and delightful. Contrasting dial pointer and knobs give more pleasing overall effect. The automatic volume control circuit keeps the reception even—no loud blasting or irritating fading. Tuning range is the standard broadcast band from 535 to 1630 KC. The set is equipped with the sensitive "Ferrite" iron rod antenna for better reception. Measures 12½" W x 7" H x 5¼" D. Operates on 105-125 volts, 60 cycles, AC or DC currents. 4 radio tubes and rectifier.

No. 1RT-55-39M (Mahogany) ....(Index T1-1729, T3-1647)............List **24.95**
No. 1RT-55391 (Ivory) ....(Index T1-2048, T3-1947)............List **29.95**
No. 1RT-55-39R (Red) ....(Index T1-2048, T3-1947)............List **29.95**
No. 1RT-55-39G (Green) ....(Index T1-2048, T3-1947)............List **29.95**

incomparable *Capehart* features

Capehart ... TABLE RADIOS
SHORT WAVE TABLE RADIO
& 3-SPEED AUTOMATIC RECORD-CHANGER

Manufactured by
CAPEHART FARNSWORTH CO.
a Division of
International Telephone & Telegraph Co.

(A) CAPEHART 5-TUBE TABLE RADIO

Eloquent testimony that "good things come in small packages," this Capehart is a trim, capable table model radio that fits in the tiniest niche while it delivers "big set" performance. 5 tubes including rectifier. Alnico special permanent magnet speaker. 5¼" high, 9⅜" wide, 5¼" deep. Operates on 105-125 volts, 50-60 cycles, AC or DC. Underwriters Laboratories approved.

No. 1RCF-2T55-E. Ebony.   No. 1RCF-2T-55-B. Brown.
(Index C1-1672, C3-1592)..................Retail **19.95**
No. 1RCF-2T55-V. Ivory.   (Index C1-1770, C3-1685)
....Retail **21.95**

(B) CAPEHART 5-TUBE TABLE RADIO

A brightly styled modern Table Radio to complement every room in the house and provide "round-the-clock" entertainment. Cabinet available in four decorator colors. 5 tubes including rectifier. Alnico special permanent magnet speaker. 5⅞" high, 9⅞" wide, 5⅝" deep. Operates on 105 to 125 volts, 50-60 cycles, AC or DC. Underwriters Laboratories approved.

No. 1RCF-3T55-E. Ebony   (Index C1-1770, C3-1685)
....Retail **21.95**
No. 1RCF-3T55-B. Brown.   (Index C1-1825, C3-1735)
....Retail **22.95**
No. 1RCF-3T55-G. Green.   No. 1RCF-3T55-V. Ivory.
(Index C1-1880, C3-1790)..................Retail **23.95**

1957 Catalog Page

(C) CAPEHART 5-TUBE TABLE RADIO

Outstanding new design that accentuates the youthful modern performance that is a true revelation of enjoyment. Cabinet 6¾" high, 9½" wide, 6" deep. Lustrefinish plastic will not fade and can be washed with soap and water. Built-in loop antenna, Alnico PM speaker with well-balanced tone quality. 5 tubes including rectifier, 105 to 125 volts, AC or DC. Underwriters Laboratories approved.

No. 1RCF-T522-V. Ivory.   No. 1RCF-T522-R. Red.
No. 1RCF-T522-SF. Silver Fox.
(Index C1-1925, C3-1830)..................Retail **24.95**

CAPEHART 5-TUBE TABLE RADIO WITH SHORT WAVE BAND 4.7 TO 18.1 MEGACYCLES

Same as above, with short wave band added for foreign reception.
No. 1RCF-T522X-V. Ivory.   No. 1RCF-T522X-R. Red.
No. 1RCF-T522X-SF. Silver Fox.
(Index C1-2756, C3-2625)..................Retail **34.95**

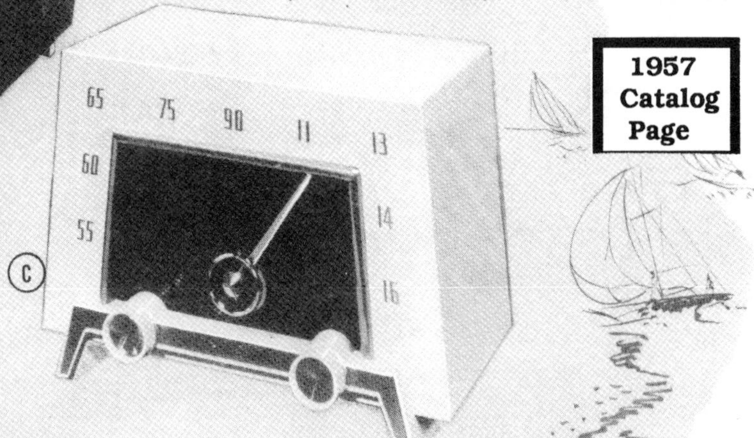

(D) THE INCOMPARABLE CAPEHART "RODEO" HIGH FIDELITY TABLE PHONOGRAPH WITH 3 SPEED AUTOMATIC RECORD CHANGER

Here is a rare find . . . a Capehart High Fidelity Table Phonograph in a beautiful Mahogany-finish cabinet that meets limited space requirements without sacrificing one bit of performance or styling. Three High fidelity speakers are perfectly balanced in this instrument for true-to-life reproduction. **Capehart High Fidelity Amplifier** . . . 4-tube "feed-back" type amplifier providing 3½ watts power output . . . more than adequate distortion-free power at normal listening levels. Full-range tone control permits adjustment to the listener's preference for highs and lows. Compensated Volume Control provides pleasant listening at all levels. **Capehart High Fidelity Speaker System** . . . Three Symphonic Tone Speakers . . . two special 6-inch speakers with oversize magnets and one Lorenz "tweeter" for wide range tonal reproduction. Electro-acoustical cross-over network channels sound into the speaker designed for its particular tonal range, contributing to the realistic tonal qualities. Entire cabinet forms special tonal chamber when lid is closed. **Capehart Record Changer** . . . plays 7-inch, 10-inch or 12-inch records at 33-1/3, 45 or 78 RPM automatically. Intermixes 10- and 12-inch records. Merely set the speed control and select the proper needle size to play all recordings. Heavy-duty motor with electrostatically-flocked, laminated turntable for quiet, efficient "wow"-free operation. Wide range, high compliance ceramic pickup with dual sapphire styli for microgroove or standard recordings. Unaffected by changes in temperature or humidity. Stabilizer arm cut out to accommodate oversize 45 rpm spindle. **Changer automatically shuts off entire instrument when last record is played.** Manual setting for children's records and home recordings. Handsome compact cabinet in genuine Mahogany veneers with hinged lid. Acoustically correct grille cloth in modern striped pattern. Height 11-5/16" width 17", depth 18⅜". 117 volts, 60 cycles, AC. (Less Stand.)
No. 1RCF-6TP45M..................(Index C1-9543, C3-9085)..................Retail **129.95**

EXCLUSIVE WROUGHT IRON STAND

For above, attractive wrought iron stand as shown. Designed exclusively for Capehart. Handy, practical shelf for storage of record albums.
No. 1RCF-BT15..................(Index C-875)..................Retail **12.00**

Capehart 3 SPEED AUTOMATIC RECORD CHANGER

ALL ITEMS ON THIS PAGE NOT INCLUDED IN OUR PREPAID FREIGHT OFFER

TRAVLER clock radios

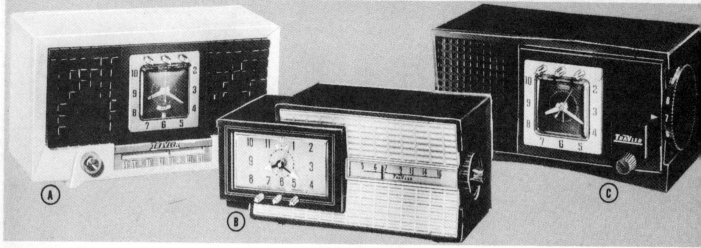

ⓐ TRAV-LER "COMPANION" 6-Tube CLOCK RADIO

Here's everything you ever wanted in a clock radio! Shows hours and minutes with easy to see luminous hands and both the day of the week and date of the month in dial face openings. Special appliance outlet for automatic off or on of radio and appliances. The attractive modern design plastic cabinet comes in choices of mahogany, Ivory and Red, Green and Yellow or Red. Easy to match the furnishings of your rooms. Has famous Telechron custom clock timer and sensitive built-in loop antenna for better reception. The distinctive bright gold knobs and dial trim add to the overall beauty of the set. Five radio tubes and rectifier. Includes two dual purpose tubes, giving 8-tube performance. Operates on 105-125V, 60 cycles, AC only.

No. 1RT-65-C45M (Mahogany).
(Index T1-3462, T3-3297)..............List **44.95**
No. 1RT-65-C45IR (Ivory and Red).
(Index T1-3777, T3-3597)..............List **49.95**
No. 1RT-65-C45R (Red).
(Index T1-3777, T3-3597)..............List **49.95**
No. 1RT-65-C45GY (Green and Yellow).
(Index T1-3777, T3-3597)..............List **49.95**
No. 1RT-65-C45B (Blonde).
(Index T1-3777, T3-3597)..............List **49.95**

ⓑ TRAV-LER 5-TUBE "MODERN DESIGN" CLOCK RADIO

Ultra modern design in gleaming plastic, available in room complimentary colors of Mahogany, Red, Ivory or Green, a rich choice of decorator colors with rich gold speaker grille and trim. A complete home servant to be treasured always. Latest advanced superheterodyne licensed under RCA and Hazeltine patents. Four radio tubes and rectifier. Includes two **dual purpose tubes giving six-tube performance.** Rectangular slide rule dial has large numbers and dial pointer for easy and relaxed tuning. Famous Telechron custom clock timer, luminous hands tell minute and hour, and dial opening tells day of week and day of month automatically. Special outlet plug for automatic timing of appliances. Turns radio or appliance off or on as desired. Measures 12½" W x 5⅞" H x 5¾" D. Operates on 105-120V, 60 cycles, AC only. Listed by Underwriters Laboratories.

No. 1RT-55-C46M (Mahogany).
(Index T1-3462, T3-3297)..............List **44.95**
No. 1RT-55-C46I (Ivory).
(Index T1-3777, T3-3597)..............List **49.95**
No. 1RT-55-C46R (Red).
(Index T1-3777, T3-3597)..............List **49.95**
No. 1RT-55-C46G (Green).
(Index T1-3777, 'T3-3597)..............List **49.95**

ⓒ TRAV-LER 5-TUBE DOUBLE DUTY CLOCK RADIO

Here's a smart modern clock radio that not only adds to the appearance of a room but can be depended upon to give the correct time of day and month, act as an alarm clock, and can turn radio or appliances on or off as desired with the special outlet built in for that purpose. The modern design is in smooth satin-like plastic. Comes in your choice of jewel-like colors: Mahogany, Red, Ivory or Green . . . all contrast beautifully with the gold finish clock dial. The luminous dial pointers are easily read and the set has a 60-minute sleep switch. The large plastic tuning knob at the end of the cabinet has easy-to-read station numbers for fine tuning. The tuning range is standard—from 535 to 1630 KC. Operates on 105-125 volts, 60 cycle, AC only.

No. 1RT-55-C42M (Mahogany).
(Index T1-2307, T3-2197)..............List **29.95**
No. 1RT-55-C42I (Ivory).
(Index T1-2517, T3-2397)..............List **32.95**
No. 1RT-55-C42R (Red).
(Index T1-2517, T3-2397)..............List **32.95**
No. 1RT-55-C42G (Green).
(Index T1-2517, T3-2397)..............List **32.95**

TRAVLER

PORTABLE BATTERY RADIOS

TRAV-LER 4-TUBE PORTABLE AC/DC AND BATTERY RADIO

This is the perfect companion for a vacation! 4 tubes plus rectifier — 2 of the tubes are dual purpose giving the receiver 6-tube performance. Tuning range from 535 to 1630 KC. The super-sensitive Ferrite iron core antenna picks up the weakest signals for long distance reception. Operates on choice of electric or battery power. Electric operation is on AC or DC current. Uses long-life batteries for most economical operation. Ideal cabinet design-presents a beautiful appearance in the home or out of doors. Practical recessed tuning dials prevent damage while carrying.

No. 1RT-5305.............(Index T1-2298, T3-2186)..............List **29.95**
No. 1RT-2-964-2-435. Battery Pack..............(Index T1-326)..............List **4.80**

TRAV-LER 4-TUBE PORTABLE "VACATIONER" BATTERY RECEIVER

A light, handy, compact receiver for any occasion! Thumb wheel tuning to balance Off/On volume control knob. Has large numbers that are easy to read and easy to tune. Equipped with 4 tubes—2 tubes are dual purpose giving 6-tube performance. Available in Red, Brown or Grey. Specify. Tuning range: 535 to 1650 KC. Gets all the standard broadcasts. Beautifully finished, lightweight plastic case with matching plastic handle and decorative rich gold finished handle and speaker trim. Measures 9" W x 2¼" D. Operates from extra long-life, 100-hour batteries. Uses one 67½-volt B battery and two 1½-volt A battery.

No. 1RT-5300.............(Index T1-1637, T3-1557)..............List **21.95**
No. 1RT-2-964-1-477. Battery Pack..............(Index T1-218)..............List **3.10**

Arvin radios

America's most Beautiful and Complete Radio Line

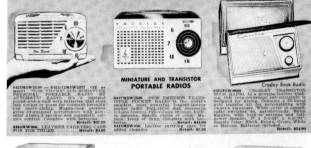

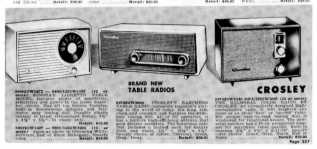

1941 MAGAZINE AD

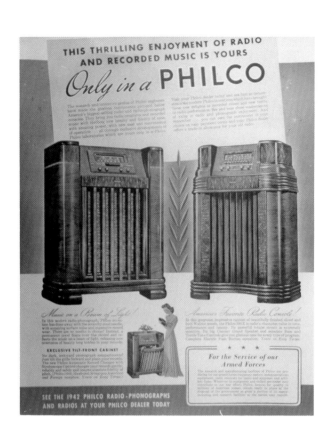

1942 MAGAZINE AD

1947 MAGAZINE AD

1948 MAGAZINE AD

1926 MAGAZINE AD

1927 MAGAZINE AD

1929 MAGAZINE AD

1929 MAGAZINE AD

1948 MAGAZINE AD

1948 MAGAZINE AD

1948 MAGAZINE AD

1949 MAGAZINE AD

1944 MAGAZINE AD

1951 MAGAZINE AD

1953 MAGAZINE AD

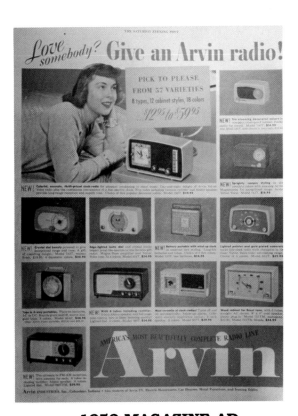

1953 MAGAZINE AD

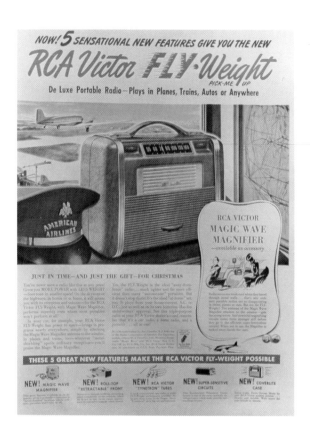

1941 MAGAZINE AD

1941 MAGAZINE AD

1939 MAGAZINE AD

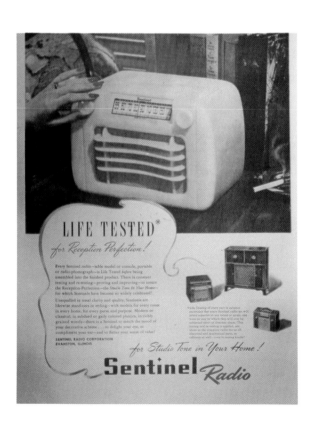

1946 MAGAZINE AD

1930 MAGAZINE AD

1936 MAGAZINE AD

1937 MAGAZINE AD

1937 MAGAZINE AD

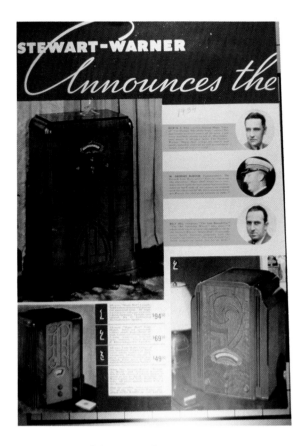

1934 MAGAZINE AD

1939 MAGAZINE AD

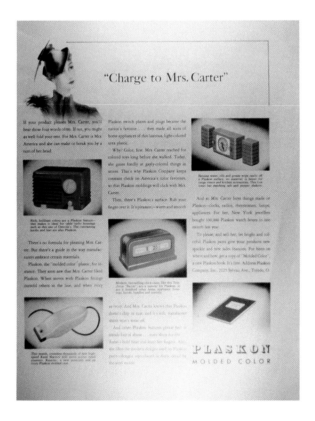

1938 MAGAZINE AD

1938 MAGAZINE AD

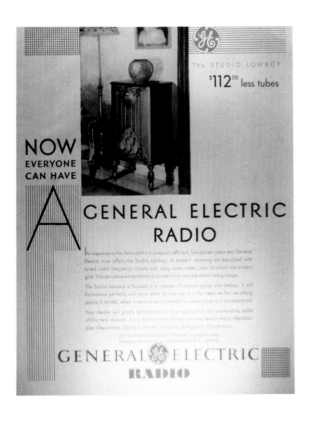

1930 MAGAZINE AD

1930 MAGAZINE AD

1936 MAGAZINE AD

1936 MAGAZINE AD

1926 MAGAZINE AD

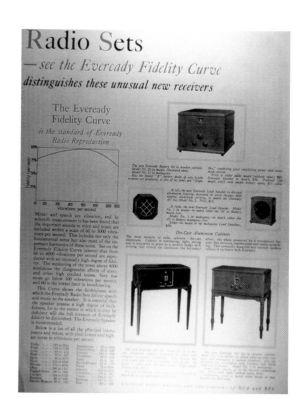

1928 MAGAZINE AD

1930 MAGAZINE AD

1930 MAGAZINE AD

1927 MAGAZINE AD

1928 MAGAZINE AD

1929 MAGAZINE AD

1929 MAGAZINE AD

1926 MAGAZINE AD

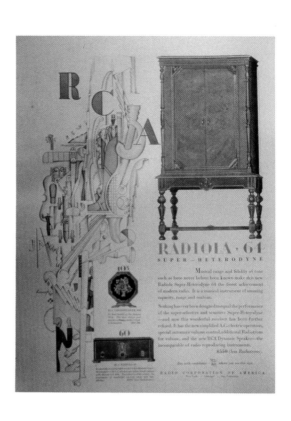

1926 MAGAZINE AD

1927 MAGAZINE AD

1929 MAGAZINE AD

Abbotwares Model #2477
1948, AM
Gerald Larsen
Elmwood Park, IL

Addison Model #2
1940, Catalin
Gary Hill
New Castle, PA

Addison Model #2F
1940, Catalin
Doug Heimstead
Fridley, MN

Addison Model #2B
1940, Catalin
Doug Heimstead
Fridley, MN

Admiral Model #990
1938, vertical or horizontal
Johnny Johnson
Denver, CO

Admiral Model #6RT42A
1947, AM & Phono
Dan Cutler
Douglas, WY

Admiral Model #5A33
1952, AM
Gerald Larsen
Elmwood Park, IL

Admiral Model #5X22
1953, AM
Gerald Larsen
Elmwood Park, IL

Admiral Model #16-D5
1941
J. H. Johnson
Greenwood, IN

Admiral Tombstone
1935, Bakelite
Gary Hill
New Castle, PA

Admiral Portable
1940, Battery
Dennis Osborne
Raleigh, NC

Admiral Model #5G32N
1957, AM
Gerald Larsen
Elmwood Park, IL

Admiral Model #5J21
1951, AM
Gerald Larsen
Elmwood Park, IL

Admiral Model #4203-B6
1942, AM
Gerald Larsen
Elmwood Park, IL

Admiral Pushbutton
1938
George Breckenridge
Gurnee, IL

Admiral Model #6Q12-N
1950, AM-FM
Gerald Larsen
Elmwood Park, IL

Advance Electric Model #4
1924
Floyd Paul
Glendale, CA

Aetna Model #635
1937
Gerald Larsen
Elmwood Park, IL

Aetna Model #300
1934, AM-SW
Gerald Larsen
Elmwood Park, IL

Aetna Model #512
1947
Gerald Larsen
Elmwood Park, IL

Aetna Model #250
1936, AM
Gerald Larsen
Elmwood Park, IL

Aetna Model #653
1937
Jim Berg
Northport, WA

Aircastle Model #711-M
1941
Gerald Larsen
Elmwood Park, IL

Air Castle Portable
1940s
Randy King
Lincoln, NE

Airite Model #4000
1938
Doug Heimstead
Fridley, MN

Air King Model #42
1935
Gerald Larsen
Elmwood Park, IL

Airline Model #62-308
1938, Movie-dial
Don Nordboe
Council Bluffs, IA

Airline Model #62-261
1937, Movie-dial
Don Nordboe
Council Bluffs, IA

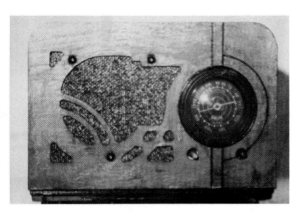

Airline Model #62-425
1936
Gerald Larsen
Elmwood Park, IL

Airline Model #74WG-2002A
1947
Spencer Doggett
Romeo, MI

Airline - Wells Gardiner
1936
Mike Feldt
Carmel, IN

Airline Model #93BR-720A
1939
Gerald Larsen
Elmwood Park, IL

Airline Model #93BR-715A
1939
Gerald Larsen
Elmwood Park, IL

Airline Movie-dial
1937
Richard's Radio of Omaha

Airline Model #05GCB-1541
1951
Johnny Johnson
Denver, CO

Airline Model #14BR-1109A
1941
Don Nordboe
Council Bluffs, IA

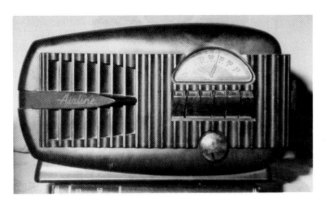

Airline Model #74BR-1507A
1947
Gerald Larsen
Elmwood Park, IL

Airline Model #93WG-602B
1940s
George Breckenridge
Gurnee, IL

Airline Model #04WG-612A
1940
Gerald Larsen
Elmwood Park, IL

Airline Model #62-288
1940s
Jay Daveler
Lansdale, PA

Airline Model #74BR-1501B
1947
Gerald Larsen
Elmwood Park, IL

Airline Model #62-445
1936
Gerald Larsen
Elmwood Park, IL

Airline Model #14BR-514B
1946
Tony's Old Ladies
Franklin Grove, IL

Airline Model #54BR-1502A
1946
Gerald Larsen
Elmwood Park, IL

Akkord Portable
1950s - W. Germany
Randy King
Lincoln, NE

Akkord Portable
1950s - W. Germany
J. Komon
W. Germany

Akkord Portable
1950s W. Germany
J. Komon
W. Germany

Alba
England
Randy King
Lincoln, NE

Alba
1930s England
Randy King
Lincoln, NE

Aldens Model #5003
1949
Gerald Larsen
Elmwood Park, IL

American Bosch Model #200B
1932
Johnny Johnson
Denver, CO

Apex Model #8A
1931
Alan Piorek
Chicago, IL

Arkay
1940s Canada
Randy King
Lincoln, NE

Arvin Model #451TL
1950
James Kroegel
Columbus, OH

Arvin Model #RE-278-2
1954 Metal
Gerald Larsen
Elmwood Park, IL

Arvin Model #422
1940 Metal
Gerald Larsen
Elmwood Park, IL

Arvin Model #444
1946 Metal
Gerald Larsen
Elmwood Park, IL

Arvin Model #253T
1950 Plastic
Gerald Larsen
Elmwood Park, IL

Arvin Model #618
1937
Gerald Larsen
Elmwood Park, IL

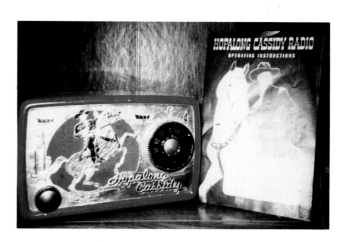

Arvin 1950
Hop-a-long Cassidy
Gary Hill
New Castle, PA

Arvin Metal
1940s
Randy King
Lincoln, NE

Atwater Kent Model #145
1934, Tombstone
James Kroegel
Columbus, OH

Atwater Kent Model #84
1931
Gary Hill
New Castle, PA

Atwater Kent Model #80
1932
Gary Hill
New Castle, PA

Atwater Kent Model #627
1933
Gary Hill

Atwater Kent Model #10
1924
Gary Hill

Atwater Kent Model #82
1931
Gary Hill
New Castle, PA

Atwater Kent Model #9C
1924
Gary Hill
New Castle, PA

Atwater Kent Model #12
1924
Gary Hill
New Castle, PA

Atwater Kent Model #30
1926
Dennis Osborne
Raleigh, NC

Atwater Kent Model #67
1929
Dan Cutler
Douglas, WY

Atwater Kent Model #20
1924
Jay Daveler
Lansdale, PA

Atwater Kent Model #52
1928, Metal
Michael Durand
Tarrytown, NY

Atwater Kent Model #84
1931
Michael Feldt

Automatic Model #613X
1946
J. E. Kendall
Fallston, MD

Atwater Kent Model #82-Q
1932
Michael Feldt
Carmel, IN

Atwater Kent Model #208
1934
Alan Piorek
Chicago, IL

Automatic Tom Thumb
1930
Peter Oppenheim
New York, NY

Balkeit Model #50A
1933
Gerald Larsen
Elmwood Park, IL

Barker Model #88
1940s, England
Randy King
Lincoln, NE

Belmont Model #6D111
1946
Gary Hill
New Castle, PA

Belmont Model #6D111
1946
Doug Heimstead
Fridley, MN

Bendix Model #301
1948
Gerald Larsen
Elmwood Park, IL

Bendix
1950s
Randy King
Lincoln, NE

Bendix Model #0526E
1946
Jay Daveler
Lansdale, PA

Benrus Clock Radio
1954
Don Nordboe
Council Bluffs, IA

Better Radio Prod.
1925
Mike Feldt
Carmel, IN

Blaupunkt
1950s, W. Germany
J. Komon
W. Germany

Blonder-Tongue Model #R986
1950s
Richard's Radio of Omaha

Brandes Model #B-15
1929
Jim Berg
Northport, WA

Bremer-Tully Kit
1925
Dennis Osborne
Raleigh, NC

Braun
1950s, W. Germany
J. Komon
W. Germany

Braun
1930s, W. Germany
Randy King
Lincoln, NE

Braun
1950s, W. Germany
Randy King

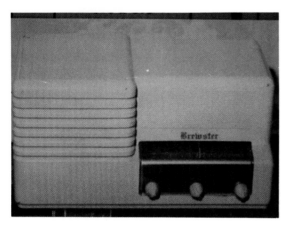

Brewster Model #9-1084
1946
Gerald Larsen
Elmwood Park, IL

Cascade
1927
Michael Feldt
Carmel, IN

Case
1926
Michael Feldt
Carmel, IN

Century Model #5-47
1932
Gerald Larsen
Elmwood Park, IL

Century "Radiolette"
1932
Gerald Larsen
Elmwood Park, IL

Chronovox Clock Radio
1931
Ross Mason
Mason City, IA

Clapp Eastham Models HR/HZ
1922
Jim Berg
Northport, WA

Climax
1939
George Breckenridge
Gurnee, IL

**Climax Ruby Supreme
1939
George Breckenridge
Gurnee, IL**

**Clinton Model #53
1934
Gerald Larsen**

**Colonial Model #654
1934
Gerald Larsen
Elmwood Park, IL**

**Colonial New World
1933
Johnny Johnson
Denver, CO**

**Continental Model #K-6
1940
Gerald Larsen
Elmwood Park, IL**

**Continental Model #1000
1945
Gerald Larsen
Elmwood Park, IL**

Columbia Model #517
1953
Jay Daveler
Lansdale, PA

Coronado Model #813
1938
Jay Daveler
Lansdale, PA

Coronado Model #550
1935
Tony's Old Ladies
Franklin Grove, IL

Coronado Model #43-8354
1948
Gerald Larsen
Elmwood Park, IL

Coronado Model #740
1936
Gerald Larsen
Elmwood Park, IL

Coronado Model #43-8190
1947
Gerald Larsen
Elmwood Park, IL

Crosley Model #B25MN
1953 Clock Radio
Don Nordboe
Council Bluffs, IA

Crosley Model #11-112U
1950s Jewelers Radio
Don Nordboe
Council Bluffs, IA

Crosley 517 "Fiver"
1937
Gerald Larsen
Elmwood Park, IL

Crosley Model #G1465
1946
Doug Heimstead
Fridley, MN

Crosley "Cub"
1931
Gerald Larsen
Elmwood Park, IL

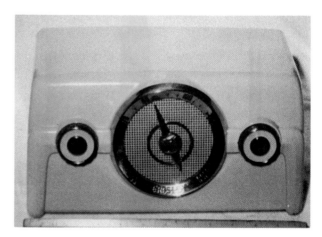

Crosley Model #10-137
1956
Gerald Larsen
Elmwood Park, IL

Crosley Model #516
1934
Gerald Larsen
Elmwood Park, IL

Crosley Model #11-103U
1951
Gerald Larsen
Elmwood Park, IL

Crosley Model #48
1931
Johnny Johnson
Denver, CO

Crosley Elf
1931
Gary Hill
New Castle, PA

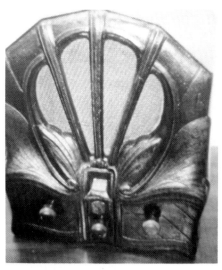

Crosley Showboy
1931
Gary Hill
New Castle, PA

Crosley Pup
1925
Gary Hill
New Castle, PA

Crosley Bullseye
1951
Gary Hill
New Castle, PA

Crosley Bullseye
1951
Gary Hill
New Castle, PA

Crosley Model #E-15-BE
1953
Richard's Radio of Omaha

Crosley Model #10-137
1950
Jay Daveler
Lansdale, PA

Crosley Model #E30-TN
1953
Spencer Doggett
Romeo, MI

Crosley Model #10-135
1950
James Kroegel
Columbus, OH

Crosley Model #66TC
1946
Jay Daveler
Lansdale, PA

Crosley Model #56TC
1947
Jay Daveler
Lansdale, PA

Crosley Model #56FA
1948
Jay Daveler
Lansdale, PA

Crosley Model #23-AR
1940s
Mark Byrd
Houston, TX

Crosley Model #627
1937
Tony's Old Ladies
Franklin Grove, IL

Crosley Model #5M3
1934, Tombstone
James Kroegel
Columbus, OH

Crosley Model #516
1934
James Kroegel

Crosley Model #635
1935
Jay Daveler
Columbus, OH

Crosley Model #59
1931, The Oracle
Carol Leeth
Anaheim, CA

Crosley Model #167
1936
Alan Piorek
Chicago, IL

Crosley Model #158
1932
James Kroegel
Columbus, OH

Crosley Model #6KA
1930s
Jay Daveler
Lansdale, PA

Crosley Model #148
1933
James Kroegel
Columbus, OH

Crosley Model #628B
1930s
James Kroegel
Columbus, OH

Crosley Model #122
1932 Super Buddy Boy
Johnny Johnson
Denver, CO

Crosley Model #125
1932
Alan Piorek
Chicago, IL

Crosley Model #706
1928 "Showbox"
Mike Feldt
Carmel, IN

Crosley Model #3R3
1924
Dennis Osborne
Raleigh, NC

Crosley Model #515
1935
James Kroegel
Columbus, OH

Crosley Model #4-29
1926
Ross Mason
Mason City, IN

Crosley Model #51
1924
James Kroegel
Columbus, OH

Crosley Super 8
1939
Randy King
Lincoln, NE

Crosley
1940s
Randy King
Lincoln, NE

Croydon Model #136E
1930s ?
Jay Daveler
Lansdale, PA

Deforest 18 Unit
1919
Ross Mason
Mason City, IA

Deforest D-6
1922
Ross Mason
Mason City, IA

Delco Model #R-1125
1937
Jim Berg
Northport, WA

Delco Model #R-1141
1937
Jay Daveler
Lansdale, PA

Detrola Pee Wee
1938
Doug Heimstead
Fridley, MN

Detrola
1938
Alan Piorek
Chicago, IL

Detrola 1939
Left/3281 - Right/302
Doug Heimstead
Fridley, MN

Detrola Model #302
1939
George Breckenridge
Gurnee, IL

Detrola Model #147
1936
Michael Feldt
Carmel, IN

Detrola Model #139
1930s
Jay Daveler
Lansdale, PA

Detrola Model #327
1930s
George Breckenridge
Gurnee, IL

Detrola
Late 1930s
Floyd Paul
Glendale, CA

Diamond T
1925
Michael Feldt
Carmel, IN

Diamond T
1926
Michael Feldt
Carmel, IN

Diamond T
1927
Michael Feldt
Carmel, IN

DKE Nazi Radio
1938
J. Komon
W. Germany

Echophone Model #S-3
1930
Alan Piorek
Chicago, IL

Echophone Model #60
1931
Gary Hill
New Castle, PA

Eagle Radio
1926, 5 tube
Dennis Osborne
Raleigh, NC

Ekco Model #A-137
1950s
Don Nordboe
Council Bluffs, IA

Elmco
1925, 5 tube
Michael Feldt
Carmel, IN

Emco Jewel
1934, Australia
Richard's Radio of Omaha

Emerson Model #440
1941
Gerald Larsen
Elmwood Park, IL

Emerson Model #45
1934, Tombstone
Michael Feldt
Carmel, IN

Emerson
1930s, Pressed Wood
Dick & Nina Davis
Baltimore, MD

Emerson Model #511
1947
Jay Daveler
Lansdale, PA

Emerson Model #BM206
1930s
Jay Daveler
Lansdale, PA

Emerson Model #DA287
1940
Jay Daveler
Lansdale, PA

Emerson Model #DB-315
1940
Jay Daveler
Lansdale, PA

Emerson Model #AX-212
1938
Richard's Radio of Omaha

Emerson
1933
Jay Daveler
Lansdale, PA

Emerson Model #574
1948
Michael Durand
Tarrytown, NY

Emerson Model #CQ271
1939
J. E. Kendall
Fallston, MD

Emerson Model #A-411
1933 - Mickey Mouse
Johnny Johnson
Denver, CO

Emerson Model #U5A
1935
Johnny Johnson
Denver, CO

Emerson Model #BA-199
1938
Jay Daveler
Lansdale, PA

Emerson Model #157
1939
Johnny Johnson
Denver, CO

Emerson Model #BM 247
1939 - Snow White & 7 Dwarfs
Johnny Johnson
Denver, CO

Emerson Model #510
1946
Jay Daveler
Lansdale, PA

Emerson Model #541
1948
Jay Daveler
Lansdale, PA

Emerson Model #520
1946, Catalin
James Kroegel
Columbus, OH

Emerson Model #602A
1949
J. E. Kendall
Fallston, MD

Emerson Model #AU 190
1939 Catalin
Johnny Johnson
Denver, CO

Emerson Model #540
1947
James Kroegel
Columbus, OH

Emerson Portable
1946
J. E. Kendall
Fallston, MD

Emerson Model #570
1949
Richard's Radio of Omaha

Emerson Model #707B
1952
Doug Heimstead
Fridley, MN

Emerson Mini Tombstone
1937
Don Nordboe
Council Bluffs, IA

Erla w/Hammond Clock
1930s
Jay Daveler
Lansdale, PA

Eveready Model #2
1928
James Kroegel
Columbus, OH

Fada Model #790
1948
Jay Daveler
Lansdale, PA

Fada Model #1005
1947
James Kroegel
Columbus, OH

Fada Model #115
1941, Catalin
Johnny Johnson
Denver, CO

Fada Model #1000
1946, Catalin
Doug Heimstead
Fridley, MN

Fada Model #260B
1937
George Breckenridge
Gurnee, IL

Fada Model #L-56
1939
Peter Oppenheim
New York, NY

Fada Model #1000
1945
Gary Hill
New Castle, PA

Fada Model #L-56
1939
Gary Hill
New Castle, PA

Fada Model #51
1931
Gary Hill
New Castle, PA

Fada Model #43
1931
Gary Hill
New Castle, PA

Farnsworth Model # ET-065
1940s
Richard Fischer
Gas City, IN

Firestone Model #4-A-12
1949
Randy King
Lincoln, NE

Firestone Clock Radio
1950s
Jay Daveler
Lansdale, PA

Firestone Model #R-320
1939
Doug Heimstead
Fridley, MN

Firestone Model #4-C-24
1952
Richard's Radio of Omaha

Freed-Eisemann Model #FE-18
1925
Jay Kinnard
Austin, TX

Freed-Eisemann Model #NR-5
1923
Dennis Osborne
Raleigh, NC

Freshman Masterpiece
1924
Jay Daveler
Lansdale, PA

Gauers Portable
Switzerland
Don Nordboe
Council Bluffs, IA

GEC
1950, England
Randy King
Lincoln, NE

General Radio Telexsa #5A
1930s
Mac's Old Time Radios
Lawndale, CA

General Electric Model #K-126
1933
Ross Mason
Mason City, IA

General Electric Model #T-41
1930
Spencer Doggett
Romeo, MI

General Electric
1933
Michael Feldt

General Electric Model #E-81
1937
Robert Breed

General Electric Model #E-105
1937
Dan Cutler
Douglas, WY

General Electric Model #J-82
1932
Gary Hill

General Electric Model #K-64
1933
Gary Hill

General Electric Model #870
1950s
Jay Daveler

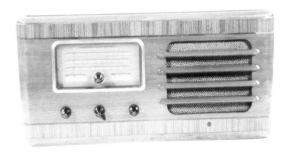

General Electric Model #F-70
1937
Jay Daveler
Lansdale, PA

General Electric Model #E-155
1936
Jim Berg
Northport, WA

General Electric Model #GB-401
1939
George Breckenridge
Gurnee, IL

General Electric Model #M-61
1934
James Kroegel
Columbus, OH

General Electric Model #H-32
1931
Dennis Osborne
Raleigh, NC

General Electric Model #K-55
1933
Jay Daveler
Lansdale, PA

General Electric Model #K-52
1933
Alan Piorek
Chicago, IL

General Electric Model #X-415
1948
Jay Daveler
Lansdale, PA

General Electric
1947
Robert Breed
San Diego, CA

General Electric Model #202
1947
James Kroegel
Columbus, OH

General Electric Model #H-500
1940
George Breckenridge
Gurnee, IL

General Electric Model #406
1950s
Gene Pupo
Spokane, WA

General Electric Model #64
1950
Spencer Doggett
Romeo, MI

General Electric Model #R-6
1930s
Don Nordboe
Council Bluffs, IA

Gulbransen Model #130
1931
Gary Hill
New Castle, PA

Gilfillan Model #84
1931
Johnny Johnson
Denver, CO

Gilfillan Model #25
1932
Johnny Johnson
Denver, CO

Gilfillan Model #8C
1935
Mac's Old Time Radios
Lawndale, CA

Gloritone Model #99
1932
Michael Feldt
Carmel, IN

Gloritone Model #3005
1930s
J. E. Kendall
Fallston, MD

Gramaphone Little Nipper
Australia
Richard's Radio of Omaha

Grantline Model #606
1947
Doug Heimstead
Fridley, MN

Greebe Model #MU1
1925
James Kroegel
Columbus, OH

Grundig Model #2120
1950s, W. Germany
Don Nordboe
Council Bluffs, IA

Grundig Majestic
1950s, W. Germany
Randy King
Lincoln, NE

Grundig Model #1083-014
1950s
Don Nordboe
Council Bluffs, IA

Grundig Model #87DX
1960s
Don Nordboe
Council Bluffs, IA

Grunow Model #4A
1932
Doug Heimstead
Fridley, MN

Grunow Model #588
1940
Don Nordboe
Council Bluffs, IA

Grunow Model #641
1935
Spencer Doggett
Romeo, MI

Grunow Model #1541
1937
Jay Daveler

Grunow Model #1291
1936
George Breckenridge
Gurnee, IL

Grunow
1933
Richad's Radio of Omaha

Guild Model #380T
1950s
Don Nordboe
Council Bluffs, IA

Guild Model #484
1956
James Kroegel
Columbus, OH

Hallicrafters Model #TW-2000
1954
Jay Daveler
Lansdale, PA

Hallicrafters Continental
1940s
Randy King
Lincoln, NE

Hallicrafters Model #TW-25
1950s
Jay Daveler
Lansdale, PA

Halston Model #65-2
1946
Gene Pupo
Spokane, WA

Hammerlund Roberts 1928
Hi-Q-29 Junior
Jim Berg
Northport, WA

Howard Green Diamond
1931
Gary Hill
New Castle, PA

Jackson-Bell Model #62
1931
Johnny Johnson
Denver, CO

Jackson-Bell Model #88
1931
Johnny Johnson
Denver, CO

JEC Model #A-244
1950s, Great Britain
Don Nordboe
Council Bluffs, IA

Jennings Coin-Op
1930
Doug Heimstead
Fridley, MN

Jewel Pinup Clock Radio
1950
Don Nordboe
Council Bluffs, IA

Kadette Model #66X
1937
Peter Oppenheim
New York, NY

Kadett Clock Radio
1939
Jay Daveler
Lansdale, PA

Kennedy Coronet
1931
Gary Hill
New Castle, PA

Kennedy XI
1924
Ross Mason
Mason City, IA

Kennedy Minerva
1929
Floyd Paul
Glendale, CA

Kennedy Model # 52A
1931
Alan Piorek
Chicago, IL

Kolster Model #6-D
1926
Dan Cutler
Douglas, WY

Kolster Brandes Model #FB10
1950, England
J. Komon
W. Germany

Lafayette
1935, 7 tubes
Michael Durand
Tarrytown, NY

Lafayette
1930, 5 tubes
Michael Durand
Tarrytown, NY

Lorenz
1945, Germany
J. Komon
W. Germany

Lucor - 77
1950s, Italy
Randy King
Lincoln, NE

Magnavox Mini Console
1930s
Don Nordboe
Council Bluffs, IA

Majestic Model #25
1932
Michael Durand
Tarrytown, NY

Majestic Model #60
1931
Michael Durand
Tarrytown, NY

Majestic Model #373
1933
Alan Piorek
Chicago, IL

Majestic Model #50
1931
James Kroegel
Columbus, OH

Majestic Model #381
1933
Michael Durand
Tarrytown, NY

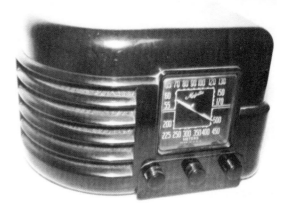

Majestic Model #7TU11
1939
George Breckenridge
Gurnee, IL

Majestic Model #15
1932
Gary Hill
New Castle, PA

Majestic Model #149
1930s
Don Nordboe
Council Bluffs, IA

Midwest Model #20-38
1938
Richard's Radio of Omaha

Midwest Model #16-37
1937
Ross Mason
Mason City, IA

Midwest Model #18-37
1937
Michael Feldt
Carmel, IN

Minerva Porta Pal
1940s
Randy King
Lincoln, NE

Mitchell Model #1250
1948
Mike Hanke
Wausau, WI

Mohawk Model R
1925
Dennis Osborne
Raleigh, NC

Motorola Model #68X-12
1940s
George Breckenridge
Gurnee, IL

Motorola Model #67X13
1949
Jay Daveler
Lansdale, PA

Motorola Model #53H
1954
George Breckenridge
Gurnee, IL

Motorola Model #52
1939
Gary Hill
New Castle, PA

Markofou
1948, Czechoslovakia
J. Komon
W. Germany

Motorola Model #51X
1941, Catalin
Gary Hill
New Castle, PA

NBC Model Galaxie 21
1950s, Poland
Don Nordboe
Council Bluffs, IA

National Union Model #G-619
1947
Jay Daveler
Lansdale, PA

Noblitt-Sparks Model #61-M
1935
Michael Feldt
Carmel, IN

No Ellen
1926
Floyd Paul
Glendale, CA

Norden-Hauck Super 10
1928
Richard's Radio of Omaha

Normende Sterling Elektra
1950s
Don Nordboe
Council Bluffs, IA

Northland
1930
Gary Hill
New Castle, PA

Olympic Opta Model #5781W
1950s, W. Germany
Ralph Michelson
Brighton, MI

Peto Scott
1954, England
Richard's Radio of Omaha

Pye Model #39J
1940s, England
Michael Feldt
Carmel, IN

Philco Model #41-290
1941
Richard Fischer
Gas City, IN

Philco Model #42-380
1942
James Kroegel
Columbus, OH

Philco Model #42-355
1942
Jay Daveler
Lansdale, PA

Philco Model #53-561
1953
Jay Daveler
Lansdale, PA

Philco Model #50-526
1950
Jay Daveler
Lansdale, PA

Philco Model #50-925
1950
David Kendall
Huntington, IN

Philco Model #53-706
1953, Lamp/Alarm Clock/Radio
Peter Oppenheim
New York, NY

Philco Model #53-562
1953
Don Nordboe
Council Bluffs, IA

Philco Model #86
1929
James Kroegel
Columbus, OH

Philco Model #49-603
1949
Richard's Radio of Omaha

Philco Model #49-909
1949
Dan Cutler
Douglas, WY

Philco Model #49-505
1949
Jay Daveler
Lansdale, PA

Philco Model #49-501
1949
Alan Piorek
Chicago, IL

Philco Model #48-206
1948
Jay Daveler
Lansdale, PA

Philco Model #48-461
1948
Jay Daveler
Lansdale, PA

Philco Model #46-420
1946
James Kroegel
Columbus, OH

Philco Model #46-1201
1946
Alan Piorek
Chicago, IL

Philco Model #41-231
1941
Doug Heimstead
Fridley, MN

Philco Model #42-321
1942
Spencer Doggett
Romeo, MI

Philco Model #16B
1933
James Kroegel
Columbus, OH

Philco Model #37-60
1937
James Kroegel
Columbus, OH

Philco Model #20B
1930
Alan Piorek
Chicago, IL

Philco Model #38-12
1938
Jay Daveler
Lansdale, PA

Philco Model #90B
1931
Alan Piorek
Chicago, IL

Philco Model #20
1930
James Kroegel
Columbus, OH

Philco Model #PT-6
1940
Jay Daveler
Lansdale, PA

Philco Model #PT44
1940
James Kroegel
Columbus, OH

Philco Model #PT3
1940
James Kroegel
Columbus, OH

Philco Model #49-902
1949
Jay Daveler
Lansdale, PA

Philco Model #116
1935
Dennis Osborne
Raleigh, NC

Philco Model #60
1934
Jim Berg
Northport, WA

Philco Model #39-80
1939
Ralph Michelson
Brighton, MI

Philco Model #37-620B
1937
James Kroegel
Columbus, OH

Philco Model #38-10
1938
David Kendall
Huntington, IN

Philco Model #610T
1936
Alan Piorek
Chicago, IL

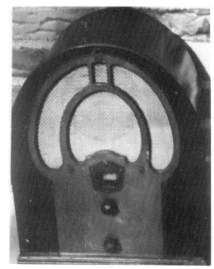

Philco Model #80B
1932
James Kroegel

Philco Model #84B
1934
Jay Daveler
Lansdale, PA

Philco Model #84
1934
Michael Feldt
Carmel, IN

Philco Model #81B
1933
James Kroegel
Columbus, OH

Philco Model #89B
1933
Dennis Osborne
Raleigh, NC

Philco Model #89B
1933
Dennis Osborne
Raleigh, NC

Philco Model #71
1932
Michael Feldt
Carmel, IN

163

Philco Model #19
1933
Alan Piorek
Chicago, IL

Philco Model #51
1932
Gary Hill
New CAstle, PA

Philco Model #90
1931
Gary Hill

Philco Model #57C
1933
Jay Daveler

Philco Model #46-1201
1946, W/78RPM Record Player
Jay Daveler
Lansdale, PA

Philco Model #3214
1940s
Jay Daveler
Lansdale, PA

Philco Model #40-185
1940
James Kroegel
Columbus, OH

Philco Model #38-690
1938
Ross Mason
Mason City, IA

Philco Model #76
1930s
Jay Daveler
Lansdale, PA

Philco Model #38-116
1938
Spencer Doggett
Romeo, MI

Philco Model #38-3
1938
Jay Daveler
Lansdale, PA

Philco Model #38-7T
1938
Richard's Radios of Omaha

Philco Model #39
1938
Alan Piorek
Chicago, IL

Peter Pan
1930s, Austrailia
Richard's Radios of Omaha

Peter Pan
1930s, Austrailia
Richard's Radios of Omaha

Peter Pan
1930s
Richard's Radios of Omaha

Packard Bell
1938
Floyd Paul
Glendale, CA

Philips Model #634
1932
J. Komon
W. Germany

Philips Model #830
1931
J. Komon
W. Germany

Philips Model #820
1931
J. Komon

Philips Model #930
1937
J. Komon
W. Germany

Philips Model #2634
1937
J. Komon
W. Germany

Philips Model #208
1942
J. Komon
W. Germany

Philips Model #944
1934
J. Komon
W. Germany

Philips Model #E5X54A
1960s
Don Nordboe
Council Bluffs, IA

Philips Model #B7 156U
1954
Richard's Radios of Omaha

Philips
1950s
Randy King
Lincoln, NE

Philips Model #BD284U
1950s
Don Nordboe
Council Bluffs, IA

Philips Georgette
1950s
J. Komon
W. Germany

Philmore Model #TR-12
1956
Richard's Radios of Omaha

Philmore Crystal Set
1940s
Gary Hill
New Castle, PA

RCA Model #101
1938
Mac's Old Time Radios
Lawndale, CA

RCA Model #R-32
1929
Dennis Osborne
Raleigh, NC

RCA Radiola 18
1927
Dave Wiggert
Onalaska, WI

RCA Radiola 17
1927
Dave Wiggert
Onalaska, WI

RCA Radiola AR-812
1924
Dave Wiggert
Onalaska, WI

RCA Radiola 60
1927
Dennis Osborne
Raleigh, NC

RCA Radiola 62
1928
Gary Hill
New Castle, PA

RCA Radiola 517
1942
Jay Daveler
Lansdale, PA

RCA Radiola 128
1934
Richard's Radios of Omaha

RCA Radiola 18
1927
Dave Wiggert
Onalaska, WI

RCA Radiola III
1924
Dennis Osborne
Raleigh, NC

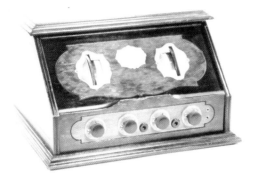

RCA Radiola 20
1925
Jay Daveler
Lansdale, PA

RCA Radiola 16
1927
Dave Wiggert
Onalaska, WI

RCA Model #7-11
1928
Dennis Osborne
Raleigh, NC

RCA Model #816K
1938
Jim Berg
Northport, WA

RCA Model #140
1933
Michael Feldt
Carmel, IN

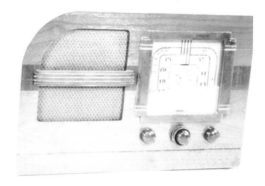

RCA Model #86T
1938
Jay Daveler

RCA Model #8BT
1938
Dave Wiggert
Onalaska, WI

RCA Model #R-5
1931
Alan Piorek
Chicago, IL

RCA Model #811K-107
1937, Electric
Jay Daveler
Lansdale, PA

RCA Model #40X57
1939
Johnny Johnson
Denver, CO

RCA Model #810T
1938
Michael Feldt
CArmel, IN

RCA Model #5T
1936
James Kroegel
Columbus, OH

RCA Model #117
1934
Peter Oppenheim
New York, NY

RCA Model #R4
1932
Gary Hill
New Castle, PA

RCA Model #86BT
1937
Don Nordboe
Council Bluffs, IA

RCA Model #9X571
1950
Jay Daveler
Lansdale, PA

RCA Model #3-BX-671
1951
Jay Daveler
Lansdale, PA

RCA Model #IT5L
1959
Mike Hanke
Wausau, WI

RCA Model #PCR-5
1950s
Johnny Johnson
Denver, CO

RCA Model #66X13
1948
James Kroegel
Columbus, OH

RCA Model #46X13
1940
Jay Daveler
Lansdale, PA

RCA Model #75X17
1948
James Kroegel
Columbus, OH

RCA Model #8BX6
1948
Mike Hanke
Wausau, WI

RCA Model #5Q55
1940
Jay Daveler
Lansdale, PA

RCA Model #66-X11
1946
Jay Daveler
Lansdale, PA

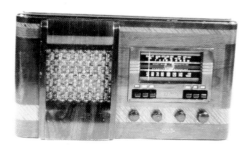

RCA Model #T62
1940
Jay Daveler
Lansdale, PA

RCA Model #15X
1940
James Kroegel
Columbus, OH

RCA Model #66X8
1946, Catalin
James Kroegel
Columbus, OH

RCA Model #8X541
1949
Jay Daveler
Lansdale, PA

RCA 1-X-57
1952
Spencer Doggett
Romeo, MI

RCA 1X53
1946
James Kroegel
Columbus, OH

RCA Model #8X682
1949
Jay Daveler
Lansdale, PA

RCA Theremin
1928
1st Synthesizer for Music
Dennis Osborne
Raleigh, NC

R. K. Radio Labs
1933 Radio Keg
Alan Piorek
Chicago, IL

Radio-Glo
Chicago's World Fair
Johnny Johnson
Denver, CO

Red Star
1954, USSR
J. Komon
W. Germany

Regentone
1940s, England
Randy King
Lincoln, NE

Remler "Scottie"
1940s
Johnny Johnson
Denver, CO

Robert Lawrence
1940s
Don Nordboe
Council Bluffs, IA

Schaub
1950s, Germany
J. Komon
W. Germany

Schneider
1950s, W. Germany
Randy King
Lincoln, NE

Scott Philharmonic Warrington
1941
Michael Feldt

Scott Philharmonic Chassis
1937, 30 tubes
Michael Feldt
Carmel, IN

Scott Phantom Deluxe
1941
Dennis Osborne
Raleigh, NC

Scott AM/FM Philharmonic
1941, 33 tubes
Michael Feldt
Carmel, IN

Scott Allwave 15
1934
Michael Feldt
Carmel, IN

Scott Allwave 15
1934 Westminster
Michael Feldt
Carmel, IN

Scott Allwave 12 Deluxe
1933
Tom Maxam
Indianapolis, IN

Setchell-Carlson
1950
Randy King
Lincoln, NE

Scott Allwave 12
1931
Michael Feldt

Sentinel Model #15516
1928
Dennis Osborne
Raleigh, NC

Siemens
1950s, W. Germany
J. Komon
W. Germany

Siemens Model #1135-W
1950s, W. Germany
Don Nordboe
Council Bluffs, IA

Silver Marshall Model #610
Late 1920s
Richard's Radios of Omaha

Silver Marshall Model #620
1928
Michael Feldt
Carmel, IN

Silvertone Model #101.432
1940s
Jay Daveler
Lansdale, PA

Silvertone Model #6356
1940s
Jay Daveler
Lansdale, PA

Silvertone Model #7025
1947
Jay Daveler
Lansdale, PA

Silvertone Model #7020
1946
Gene Pupo
Spokane, WA

Silvertone Model #7140
1930s
Spencer Doggett
Romeo, MI

Silvertone Model #109.216
1939
Jay Kinnard
Austin, TX

Silvertone Model #1628
1934
Carol Leeth
Anaheim, CA

Silvertone Model #4485
1937
Jay Daveler
Lansdale, PA

Silvertone #4586
1930s
Jay Daveler
Lansdale, PA

Silvertone Model #1412
1930s
Jay Daveler
Lansdale, PA

Silvertone
1936
Michael Feldt

Silvertone
1933
Michael Feldt
Carmel, IN

Silvertone Model #1807
1934
James Kroegel
Columbus, OH

Silvertone Model #115
1929
Carol Leeth
Anaheim, CA

Sky Rover
1931, 7 tubes
Michael Durand
Tarrytown, NY

Sonora Excellence
1940s, France
J. Komon
W. Germany

Sonora Sonorette
1940s, France
J. Komon
W. Germany

Sparton Model #620M
1930s
Jay Daveler
Lansdale, PA

Sparton
1940s
Randy King
Lincoln, NE

Sparton Model #842-5X
1942
Michael Feldt
Carmel, IN

Sparton Model #558
1937 Mirror
Johnny Johnson
Denver, CO

Sparton Model #1486
1936
Doug Heimstead
Fridley, MN

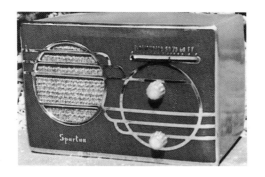

Sparton Model #500C
1939
Doug Heimstead
Fridley, MN

Sparton Model #500C
1939
Doug Heimstead
Fridley, MN

Sparton Model #N-61
1933
Gene Pupo
Spokane, WA

Sparton Model #62
1927
Dennis Osborne
Raleigh, NC

STC Standard Telephones
1934, Orchestrelle Co., Melborn
Richard's Radios of Omaha

Steinite Model #262
1928
Dennis Osborne
Raleigh, NC

Stewart Warner Model #01-521
1930s
Don Nordboe
Council Bluffs, IA

Stewart Warner Model #B51T2
1949
J. E. Kendall
Fallston, MD

Stewart Warner Model #B51T1
1949
George Breckenridge
Gurnee, IL

Stewart Warner
1933
Doug Heimstead
Fridley, MN

Stewart Warner
1935
Robert Breed
San Diego, CA

Stewart Warner Model #R123
1935
Richard's Radios of Omaha

Stewart Warner Model #91-513
1938
Johnny Johnson
Denver, CO

Stewart Warner Dionne Quints
1939
George Breckenridge
Gurnee, IL

Stewart Warner Model #525
1927
Jay Daveler
Lansdale, PA

Stewart Warner Model #206
1940s
Jay Daveler
Lansdale, PA

Stewart Warner "Apartment"
1931
Gary Hill
New Castle, PA

Stromberg Carlson Model #69
1934
Jay Daveler

Stromberg Carlson Model #255
1937
George Breckenridge
Gurnee, IL

Stromberg Carlson Model #82
1935
Jim Berg
Northport, WA

Stromberg Carlson Model #1400-H
1949
Jay Daveler
Lansdale, PA

Stromberg Carlson Model #1101-H
1947
Tony's Old Ladies
Franklin Grove, IL

Stromberg Carlson Model #225H
1936
Doug Heimstead
Fridley, MN

Stromberg Carlson Model #AWB-8
1950s
Jay Daveler
Lansdale, PA

Tailsman Model #307V 1
1940s, Germany
Gene Pupo
Spokane, WA

Tailsman Model #308V
1940s, Germany
Gene Pupo
Spokane, WA

Telefunken Model #31W
1927, Germany
Richard's Radios of Omaha

Telefunken Kavalier
1950
Richard's Radios of Omaha

Telefunken Bajazzo
1950s, Germany
Randy King
Lincoln, NE

Telefunken - Opus 6
1950s, Germany
Randy King
Lincoln, NE

Telefunken Model #340
1932, "Cat Head"
J. Komon
W. Germany

Telefunken Model #500
1933, Germany
J. Komon
W. Germany

Telefunken Model #33
1931, Germany
J. Komon
W. Germany

Telex Hospital Coin-Op
Model #T-6, 1950s
Don Nordboe
Council Bluffs, IA

Tesla Minor
1950s
J. Komon
W. Germany

Tesla Model #306
1950s, Czechoslovakia
J. Komon
W. Germany

Tesla Model #308
1950s, Czechoslovakia
J. Komon
W. Germany

Tesla Rekreant
1955, Czechoslovakia
J. Komon
W. Germany

Tilman Model #T-18
1925
Michael Feldt
Carmel, IN

Trav-ler Model #635-M
1938, "Blackhawk"
Gerald Larsen
Elmwood Park, IL

Trav-ler Model #5300
1954
Randy King
Lincoln, NE

**Truetone Superhet
1930s
Richard's Radios of Omaha**

**Truetone Model #D2210
1941
Richard's Radios of Omaha**

**Truetone Model #D1612
1947
James Kroegel
Columbus, OH**

**Turist
1958, USSR
J. Komon
W. Germany**

**U. S. Radio Model 9A
1930s
Don Nordboe
Council Bluffs, IA**

**Viz Model #R51
1947
Gene Pupo
Spokane, WA**

Watterson Texas Centennial
1936
Gil Barborak
Austin, TX

Westinghouse RC
1922
Dennis Osborne
Raleigh, NC

Westinghouse Model #H-359TF
1953
Richard's Radios of Omaha

Westinghouse Model #H397T5
1955
Don Nordboe
Council Bluffs, IA

Westinghouse Model #H398T5
1955
Don Nordboe
Council Bluffs, IA

Westinghouse Model #H124
1949
Michael Durand
Tarrytown, NY

Westinghouse Model #H-126
1948
James Kroegel
Columbus, OH

Westinghouse Model #501
1948
Gene Pupo
Spokane, WA

Westinghouse Model #U2153
1948
Richard's Radios of Omaha

Westinghouse Model #WR-8
1931
Dennis Osborne
Raleigh, NC

Wilcox-Gay Model #7A5
1936
Richard's Radios of Omaha

Wilson
1936, 6 tubes
Michael Feldt
Carmel, IN

Wilson
1934, 5 tubes
Michael Feldt

Wurlitzer Model #SW-88
1930s
Don Nordboe
Council Bluffs, IA

Zaney Gill
1931
Johnny Johnson
Denver, CO

Zenith Model #15U269
1940s
Jay Daveler
Lansdale, PA

Zenith Model #9-S-54
1930s
Jay Daveler

Zenith Model #8-D-510
1940s
Jay Daveler

Zenith Model #6D269
1940s
Jay Daveler
Lansdale, PA

Zenith Model #6D2615
1940s
Jay Daveler
Lansdale, PA

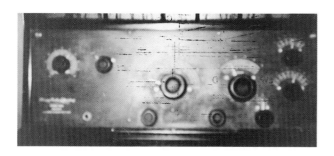

Zenith Model #3R
1923
Jim Berg
Northport, WA

Zenith Model #6S527
1930s
James Kroegel
Columbus, OH

Zenith Model #805
1934
Alan Piorek
Chicago, IL

Zenith Model #835
1933
Doug Heimstead
Fridley, MN

Zenith Model #6S229
1937
Jay Daveler
Lansdale, PA

Zenith Model #4U31
1935
James Kroegel
Columbus, OH

Zenith Model #6B129
1936
James Kroegel
Columbus, OH

Zenith Model #6D137
1938
George Breckenridge
Gurnee, IL

Zenith Model #6S203
1938
Robert Breed
San Diego, CA

Zenitithh "Zenette"
1931
Robert Breed
San Diego, CA

Zenith "Zenette"
1931
Robert Breed

Zenith Model #5S327
1939
Robert Breed
San Diego, CA

Zenith Model #5S29
1936
Robert Breed
San Diego, CA

Zenith Model #5R312
1938
J. E. Kendall
Fallston, MD

Zenith Model #9S244
1937 Chairside
James Kroegel
Columbus, OH

Zenith Model #5S319
1937
Jay Daveler
Lansdale, PA

Zenith Model #5S228
1938
Robert Breed
San Diego, CA

Zenith Model #9S30
1936
Robert Breed
San Diego, CA

Zenith Model #6D315U
1938
George Breckenridge
Gurnee, IL

Zenith Model #H664
1951
J. E. Kendall
Fallston, MD

Zenith Model #6D2615
1942
Spencer Doggett
Romeo, MI

Zenith Model #7S933
1940s
Jay Daveler
Lansdale, PA

Zenith Model #5D011W
1946
Jay Daveler
Lansdale, PA

Zenith Model #8S463
1939
Dan Cutler
Douglas, WY

Zenith Model #8S154
1936
George Breckenridge
Gurnee, IL

Zenith Model #7S363
1939
Jay Daveler
Lansdale, PA

Zenith Model #880
1934
Jim Berg
Northport, WA

Zenith Model #72
1930
Ross Mason
Mason City, IA

Zenith #12S265
1938
Dale Brown
Costa Mesa, CA

Zenith #12A58
1936
Michael Feldt
Carmel, IN

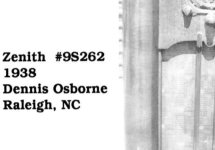

Zenith #9S262
1938
Dennis Osborne
Raleigh, NC

Zenith #9S367
1939
Dennis Osborne
Raleigh, NC

Zenith #6S152
1936
James Kroegel
Columbus, OH

Zenith #10S464
1940
Dennis Osborne
Raleigh, NC

Zenith Model #5808
1940
Michael Feldt
Carmel, IN

Zenith Model #10S668
1942
Mark Byrd
Houston, TX

Zenith Model #10S470
1940
Johnny Johnson
Denver, CO

Zenith Model #9S262
1938
Doug Heimstead
Fridley, MN

Zenith Model #705
1930s
Don Nordboe
Council Bluffs, IA

Zenith Model #7S634
1941
Doug Heimstead
Fridley, MN

Zenith Spinet Piano
1941
George Breckenridge
Gurnee, IL

Zenith Model #6D526
1941
Gene Pupo
Spokane, WA

Zenith Model #6D525
1941
Jay Daveler
Lansdale, PA

Zenith Model #6T41
1940s
Jay Daveler
Lansdale, PA

Zenith Model #G500
1940s
Jay Daveler
Lansdale, PA

Zenith Royal 3000-1
1940s
Jay Daveler
Lansdale, PA

Zenith R-7000
1940s
Jay Daveler
Lansdale, PA

Zenith Royal 1000
1940s
Jay Daveler
Lansdale, PA

Zenith RD 7000Y
1940s
Jay Daveler
Lansdale, PA

Zenith Model #8G005YT
1946
Dennis Osborne
Raleigh, NC

Zenith Model # 12S232 - 1938
Used in the Opening Credits of the "Waltons" Series
Robert Breed, San Diego, CA

Coca Cooler Radio
1949
Gary Hill
New Castle, PA

Table Lamp Radio
Jay Daveler

Bottle Radio
1940s
Jay Daveler
Lansdale, PA

Universal Model #171636
1960, Mike Radio
J. H. Johnson
Greenwood, IN

Tower Speaker
1926
Gary Hill

Utah Speaker
1926
Gary Hill

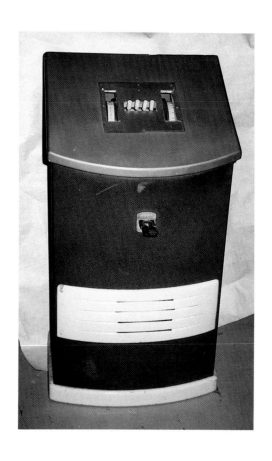

Top Left: Console 10¢ Coin-Op, 1933

Top Right: Console 25¢ Coin-Op, 1935

Bottom: Artist Designed, Note Plastic Soldiers
1960, Made in Los Angeles

All three owned by: Mac's Old Time Radios
Lawndale, CA

Actual Postcards

His Master's V-ice!

Actual Postcards

„Dé nieuws berichten"

CHRISTMAS GREETINGS

A COUPLE LIKE YOU OUGHT TO HAVE A **WIRELESS SET.** YOU COULD CANOODLE BETWEEN THE ITEMS WITHOUT AN AUDIENCE LOOKING ON!

Actual Postcards

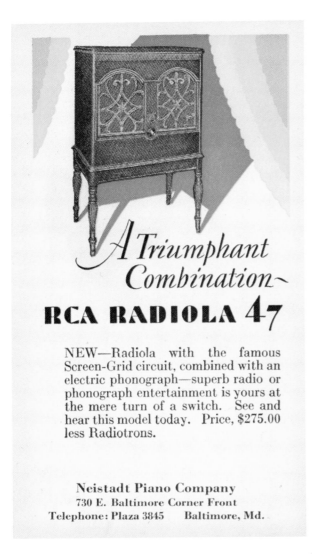

A Triumphant Combination

RCA RADIOLA 47

NEW—Radiola with the famous Screen-Grid circuit, combined with an electric phonograph—superb radio or phonograph entertainment is yours at the mere turn of a switch. See and hear this model today. Price, $275.00 less Radiotrons.

Neistadt Piano Company
730 E. Baltimore Corner Front
Telephone: Plaza 3845   Baltimore, Md.

To FIT COZILY *into a* CORNER

RCA RADIOLA 44
LOUDSPEAKER 103

THE new AC Screen-Grid Radiola 44. Fine selectivity and sensitivity. Two-in-one control. $110.00 less Radiotrons. RCA Loudspeaker 103, $22.50

Neistadt Piano Company
730 E. Baltimore Corner Front
Telephone: Plaza 3845   Baltimore, Md.

PRICE GUIDE

Page 1
Top Left #62 - $150
Top Left #61-H - $75
Top Right - $125
Bottom Left - $100 - 150
Bottom Right - $210

Page 2
Reference Only

Page 3
$100

Page 4
$200

Page 5
$250

Page 6
Left - $200
Right - $80

Page 7
$100

Page 8
Left - $125
Right - $150

Page 9
Top Left - $300
Top Right - $50-150
Bottom Left - $250
Bottom Right - $300

Page 10
Top Left - $75-150
Top Right - $125
Bottom Left - $50
Bottom Right - $75

Page 11
Top Left - $150-250
Top Right - Ref. only
Bottom Left - Ref. only
Bottom Right - Ref. only

Page 12
Top Left Arbiter - $200
Top Right - Ref. only
Bottom Left - Ref. only
Bottom Right - $150

Page 13
Top Left - $150
Top Right - $125
Middle Left - $150
Middle Right - $125
Bottom Left - $125
Bottom Right - $110

Page 14
AK35 - $65
Radiola 25 - $200
Freshman A - $100
Freshman B - $100
SC 523 - $125
Ware T - $265
Radiola 20 - $175

Page 15
Top Left - $150
Top Right - $250
Bottom Left - $150
Bottom Right - $175

Page 16
Top Left - $100
Top Right - $200
Bottom - $100

Page 17
$100 - 150

Page 18
$150 - 200

Page 19
$75 - 150

Page 20
$100 - 150

Page 21
Top Left - $125 - 250
Top Right - $150
Bottom Left - $100 - 200
Bottom Right - $125 - 200

Page 22
All on page - $125 - $175

Page 23
$50 - 100

Page 24
$100 - 150

Page 25
Top - $100
Bottom - $100

Page 26
$100 - 150

Page 27
$50 - 75

Page 28
Top - $75
Bottom - $100

Page 29
$40 - 80

Page 30
Top Left - $60
Top Right - $50
Bottom Left - $50 -100
Bottom Right - $50

Page 31
$100 - 200

Page 32
Top - $40
Middle - $100
Bottom - $50

Page 33
Top Left - $200
Top Right - $150
Bottom Left - $150
Bottom Right - Ref. only

Page 34
Top Left - $50
Top Right - $50 - 75
Bottom Left - $50
Bottom Right - $60

Page 35
Top Left - $100
Top Right - $50 - 100
Bottom Left - $50
Bottom Right - $50 - 100

Page 36
Top Left - $200
Top Right - $50 - 100
Bottom Left - $150
Bottom Right - $150

PRICE GUIDE

Page 37
Top Left - $125
Top Right - $50
Bottom Left - $50 -100
Bottom Right - $150

Page 38
Top Left - $150
Top Right - $50 - 100
Bottom Left - $50 - 200
Bottom Right - $50 - 100

Page 39
$50 - 75

Page 40
$50 - 200

Page 41
$75 - 200

Page 42
$50 - 75

Page 43
$50 - 75

Page 44
$50 - 75

Page 45
$50 - 200

Page 46
Top - $125 - 175
Bottom - $50 - 75

Page 47
$50 - 75

Page 48
Top - $100 - 150
Bottom - $50 - 75

Page 49
$50 - 75

Page 50
$50 - 75

Page 51
$50 - 75

Page 52
$50 - 75

Page 53
Top Left - $100 - 200
Top Right - $50 - 75
Bottom Left - $150 - 300
Bottom Right - $50 - 75

Page 54
Top - $50 - 200
Bottom - $40 - 75

Page 55
$50 - 100

Page 56
Top Left - $50 - 75
Top Right - $50 and $2,000
Bottom Left - $50 - 150
Bottom Right - Ref. only

Page 57
Bottom Right - $50 - 75
All others - $100 - 200

Page 58
Top - $125
Bottom - $50

Page 59
$50 - 75

Page 60
Top - $50
Bottom - $150

Page 61
$50 - 100

Page 62
$40 - 60

Page 63
$40 - 60

Page 64
$50 - 75

Page 65
$70 - 75

Page 66
$40 - 60

Page 67
$40 - 60

Page 68
Top - $75
Bottom - $150

Page 69
$50 - 100

Page 70
$100 - 200

Page 71
Top Left - $150
Top Right - $50 - 75
Bottom Left - $150
Bottom Right - $50 - 75

Page 72
$50 - 100

Page 73
Top Left - $150
Top Right - $50 - 75
Bottom Left - $50 - 75
Bottom Right - $50 and $1,500

Page 74
Top Left - $50 - 75
Top Right - $50 - 75
Bottom Left - $50 - 75
Bottom Right - $125

Page 75
$50 - 75

Page 76
$50 - 75

Page 77
$50 - 75

Page 78
$50 - 75

Page 79
Top - $1,800
Middle - $50
Bottom - $50 - 75

Page 80
$40 - 60

Page 81
$30 - 40

PRICE GUIDE

Page 82
$30 - 40

Page 83
$30 - 40

Page 84
$30 - 50

Page 85
$30 - 40

Page 86
$30 - 50

Page 87
$30 - 50

Page 88
$30 - 50

Page 89
$30 - 40

Page 90
$30 - 40

Page 91

Top Left -	$50
Top Right -	$30 - 50
Bottom Left -	$50
Bottom Right -	$30 - 50

Page 92

Top Right -	$125 - 175
All others -	$50 - 75

Page 93
Reference only

Page 94

Top -	$100 - 175
Bottom -	$50 - 75

Page 95
$50 - 75

Page 96
$50 - 75

Page 97
$150 - 250

Page 98
Reference only

Page 99
Reference only

Page 100
Reference only

Page 101
Reference only

Page 102
Reference only

Page 103

Top Left -	$225
Top Right -	$500 +
Middle Left -	$500 +
Middle Right -	$500 +
Bottom Left -	$125
Bottom Right -	$35

Page 104

Top Left -	$40
Top Right -	$40
Middle Left -	$70
Middle Right -	$90
Bottom Left -	$45
Bottom Right -	$40

Page 105

Top Left -	$40
Top Right -	$55
Middle Left -	$50
Middle Right -	$40
Bottom Left -	$200
Bottom Right -	$95

Page 106

Top Left -	$75
Top Right -	$30
Middle Left -	$45
Middle Right -	$80
Bottom Left -	$40
Bottom Right -	$35

Page 107

Top Left -	$200
Top Right -	$35
Middle Left -	$250 +
Middle Right -	$300 +
Bottom Left -	$50
Bottom Right -	$40

Page 108

Top Left -	$150
Top Right -	$40
Middle Left -	$70
Middle Right -	$175
Bottom Left -	$175
Bottom Right -	$70

Page 109

Top Left -	$90
Top Right -	$80
Middle Left -	$50
Middle Right -	$65
Bottom Left -	$40
Bottom Right -	$50

Page 110

Top Left -	$90
Top Right -	$40
Middle Left -	$50
Middle Right -	$75
Bottom Left -	$75
Bottom Right -	$50

Page 111

Top Left -	$125
Top Right -	$40
Middle Left -	$150
Middle Right -	$200
Bottom Left -	$60
Bottom Right -	$45

Page 112

Top Left -	$45
Top Right -	$45
Middle Left -	$45
Middle Right -	$35
Bottom Left -	$55
Bottom Right -	$500 +

Page 113

Top Left -	$40
Top Right -	$175
Middle Left -	$400 +
Middle Right -	$400 +
Bottom Left -	$550 +
Bottom Right -	$850 +

Page 114

Top Left -	$400 +
Top Right -	$1,100 +
Middle Left -	$1,300 +
Middle Right -	$200
Bottom Left -	$80
Bottom Right -	$120

PRICE GUIDE

Page 115
Top Left -	$200
Top Right -	$375
Middle Left -	$325
Middle Right -	$250
Bottom Left -	$65
Bottom Right -	$125

Page 116
Top Left -	$40
Top Right -	$75
Middle Left -	$185
Middle Right -	$185
Bottom Left -	$40
Bottom Right -	$40

Page 117
Top Left -	$65
Top Right -	$100
Middle Left -	$150
Middle Right -	$100
Bottom Left -	$40
Bottom Right -	$225

Page 118
Top Left -	$125
Top Right -	$100
Middle Left -	$200
Middle Right -	$225
Bottom Left -	$50
Bottom Right -	$125

Page 119
Top Left -	$100
Top Right -	$175
Middle Left -	$150
Middle Right -	$500 +
Bottom Left -	$600 +
Bottom Right -	$125

Page 120
Top Left -	$175
Top Right -	$35
Middle Left -	$40
Middle Right -	$600 +
Bottom Left -	$40
Bottom Right -	$75

Page 121
Top Left -	$45
Top Right -	$40
Middle Left -	$45
Middle Right -	$35
Bottom Left -	$45
Bottom Right -	$70

Page 122
Top Left -	$100
Top Right -	$110
Middle Left -	$55
Middle Right -	$1,500 +
Bottom Left -	$125
Bottom Right -	$75

Page 123
Top Left -	$125
Top Right -	$90
Middle Left -	$300 +
Middle Right -	$500 +
Bottom Left -	$500 +
Bottom Right -	$325 +

Page 124
Top Left -	$125
Top Right -	$125
Middle Left -	$75
Middle Right -	$75
Bottom Left -	$40
Bottom Right -	$75

Page 125
Top Left -	$70
Top Right -	$45
Middle Left -	$40
Middle Right -	$60
Bottom Left -	$60
Bottom Right -	$100

Page 126
Top Left -	$95
Top Right -	$100
Middle Left -	$600 +
Middle Right -	$165
Bottom Left -	$200
Bottom Right -	$125

Page 127
Top Left -	$170
Top Right -	$45
Middle Left -	$200
Middle Right -	$150
Bottom Left -	$125
Bottom Right -	$110

Page 128
Top Left -	$90
Top Right -	$125
Middle Left -	$125
Middle Right -	$65
Bottom Left -	$75
Bottom Right -	$100

Page 129
Top Left -	$1,700 +
Top Right -	$1,400 +
Middle Left -	$50
Middle Right -	$45
Bottom Left -	$200 +
Bottom Right -	$85

Page 130
Top Left -	$250 ea.
Top Right -	$250
Middle Left -	$125
Middle Right -	$60
Bottom Left -	$60
Bottom Right -	$65

Page 131
Top Left -	$100
Top Right -	$100
Middle Left -	$100
Middle Right -	$225
Bottom Left -	$250 +
Bottom Right -	$350 +

Page 132
Top Left -	$125
Top Right -	$140
Middle Left -	$100
Middle Right -	$200
Bottom Left -	$45
Bottom Right -	$150

Page 133
Top Left -	$150
Top Right -	$45
Middle Left -	$45
Middle Right -	$50
Bottom Left -	$65
Bottom Right -	$125

Page 134
Top Left -	$75
Top Right -	$150
Middle Left -	$85
Middle Right -	$1,000 +
Bottom Left -	$230
Bottom Right -	$110

Page 135
Top Left -	$60
Top Right -	$1,000 +
Middle Left -	$40
Middle Right -	$40
Bottom Left -	$225
Bottom Right -	$75

PRICE GUIDE

Page 136
Top Left - $1,000 +
Top Right - $50
Middle Left - $75
Middle Right - $75
Bottom Left - $125
Bottom Right - $100

Page 137
Top Left - $350
Top Right - $125
Middle Left - $100
Middle Right - $175
Bottom Left - $600 +
Bottom Right - $650 +

Page 138
Top Left - $100
Top Right - $750 +
Middle Left - $2,000 +
Middle Right - $2,000 +
Bottom Left - $275 +
Bottom Right - $450 +

Page 139
Top Left - $85
Top Right - $60
Middle Left - $50
Middle Right - $100
Bottom Left - $35
Bottom Right - $75

Page 140
Top Left - $130
Top Right - $130
Middle Left - $70
Middle Right - $100
Bottom Left - $300 +
Bottom Right - $200

Page 141
Top Left - $125
Top Right - $125
Middle Left - $125
Middle Right - $150
Bottom Left - $300 +
Bottom Right - $275 +

Page 142
Top Left - $50
Top Right - $70
Middle Left - $125
Middle Right - $75
Bottom Left - $175
Bottom Right - $150

Page 143
Top Left - $125
Top Right - $200
Middle Left - $100
Middle Right - $100
Bottom Left - $50
Bottom Right - $100

Page 144
Top Left - $65
Top Right - $40
Middle Left - $200
Middle Right - $300 +
Bottom Left - $200
Bottom Right - $200

Page 145
Top Left - $300 +
Top Right - $250
Middle Left - $100
Middle Right - $225
Bottom Left - $125
Bottom Right - $250

Page 146
Top Left - $100
Top Right - $50
Middle Left - $60
Middle Right - $75
Bottom Left - $150
Bottom Right - $150

Page 147
Top Left - $150
Top Right - $225
Middle Left - $225
Middle Right - $100
Bottom Left - $150
Bottom Right - $100

Page 148
Top Left - $100
Top Right - $50
Middle Left - $50
Middle Right - $60
Bottom Left - $125
Bottom Right - $350 +

Page 149
Top Left - $275
Top Right - $250
Middle Left - $75
Middle Right - $400 +
Bottom Left - $75
Bottom Right - $65

Page 150
Top Left - $175
Top Right - $300
Middle Left - $400
Middle Right - $100
Bottom Left - $100
Bottom Right - $100

Page 151
Top Left - $125
Top Right - $175
Middle Left - $175
Middle Right - $125
Bottom Left - $60
Bottom Right - $200

Page 152
Top Left - $250
Top Right - $250
Middle Left - $125
Middle Right - $300
Bottom Left - $250
Bottom Right - $75

Page 153
Top Left - $150
Top Right - $125
Middle Left - $500
Middle Right - $300
Bottom Left - $300
Bottom Right - $40

Page 154
Top Left - $100
Top Right - $150
Middle Left - $50
Middle Right - $40
Bottom Left - $60
Bottom Right - $1,200 +

Page 155
Top Left - $150
Top Right - $2,500 +
Middle Left - $125
Middle Right - $40
Bottom Left - $125
Bottom Right - $200 +

Page 156
Top Left - $2,000 +
Top Right - $100
Middle Left - $300 +
Middle Right - $125
Bottom Left - $750 +
Bottom Right - $150

PRICE GUIDE

Page 157
Top Left -	$175
Top Right -	$150
Middle Left -	$75
Middle Right -	$35
Bottom Left -	$45
Bottom Right -	$35

Page 158
Top Left -	$125
Top Right -	$35
Middle Left -	$150
Middle Right -	$50
Bottom Left -	$40
Bottom Right -	$50

Page 159
Top Left -	$85
Top Right -	$35
Middle Left -	$35
Middle Right -	$50
Bottom Left -	$75
Bottom Right -	$85

Page 160
Top Left -	$45
Top Right -	$240
Middle Left -	$125
Middle Right -	$225
Bottom Left -	$50
Bottom Right -	$400

Page 161
Top Left -	$225
Top Right -	$50
Middle Left -	$45
Middle Right -	$45
Bottom Left -	$35
Bottom Right -	$125

Page 162
Top Left -	$115
Top Right -	$100
Middle Left -	$130
Middle Right -	$100
Bottom Left -	$125
Bottom Right -	$165

Page 163
Top Left -	$165
Top Right -	$175
Middle Left -	$165
Middle Right -	$150
Bottom Left -	$150
Bottom Right -	$200

Page 164
Top Left -	$150
Top Right -	$400 +
Middle Left -	$500 +
Middle Right -	$80
Bottom Left -	$75
Bottom Right -	$45

Page 165
Top Left -	$135
Top Right -	$250
Bottom Left -	$125
Bottom Right -	$200

Page 166
Top Left -	$125
Top Right -	$75
Middle Left -	$45
Middle Right -	$500 +
Bottom Left -	$250
Bottom Right -	$250

Page 167
Top Left -	$70
Top Right -	$150
Middle Left -	$175
Middle Right -	$175
Bottom Left -	$150
Bottom Right -	$150

Page 168
Top Left -	$125
Top Right -	$150
Middle Left -	$125
Middle Right -	$100
Bottom Left -	$65
Bottom Right -	$65

Page 169
Top Left -	$75
Top Right -	$100
Middle Left -	$40
Middle Right -	$200
Bottom Left -	$150
Bottom Right -	$110

Page 170
Top Left -	$100
Top Right -	$210
Middle Left -	$120
Middle Right -	$300 +
Bottom Left -	$35
Bottom Right -	$285

Page 171
Top Left -	$110
Top Right -	$110
Middle Left -	$225
Middle Right -	$110
Bottom Left -	$200
Bottom Right -	$210

Page 172
Top Left -	$275
Top Right -	$80
Middle Left -	$175
Middle Right -	$200
Bottom Left -	$150
Bottom Right -	$800 +

Page 173
Top Left -	$150
Top Right -	$105
Middle Left -	$150
Middle Right -	$325
Bottom Left -	$75
Bottom Right -	$50

Page 174
Top Left -	$125
Top Right -	$30
Middle Left -	$200
Middle Right -	$50
Bottom Left -	$40
Bottom Right -	$165

Page 175
Top Left -	$45
Top Right -	$50
Middle Left -	$50
Middle Right -	$75
Bottom Left -	$40
Bottom Right -	$600 +

Page 176
Top Left -	$30
Top Right -	$40
Middle Left -	$40
Middle Right -	$40
Bottom Left -	Rare
Bottom Right -	$375

Page 177
Top Left -	$300 +
Top Right -	Rare
Middle Left -	$75
Middle Right -	$100
Bottom Left -	$70
Bottom Right -	$100

PRICE GUIDE

Page 178
Top Left -	$50
Top Right -	$1,800
Middle Left -	$600
Middle Right -	$525
Bottom Left -	$600
Bottom Right -	$300

Page 179
Top Left -	$325
Top Right -	$300
Middle Left -	$300
Middle Right -	$175
Bottom Left -	$75
Bottom Right -	$150

Page 180
Top Left -	$100
Top Right -	$300
Middle Left -	$150
Middle Right -	$75
Bottom Left -	$75
Bottom Right -	$65

Page 181
Top Left -	$75
Top Right -	$130
Middle Left -	$80
Middle Right -	$250
Bottom Left -	$150
Bottom Right -	$150

Page 182
Top Left -	$110
Top Right -	$75
Middle Left -	$75
Middle Right -	$100
Bottom Left -	$250
Bottom Right -	$275

Page 183
Top Left -	$150
Top Right -	$150
Middle Left -	$75
Middle Right -	$50
Bottom Left -	$100
Bottom Right -	$2,000 +

Page 184
Top Left -	$1,000 +
Top Right -	$2,200 +
Middle Left -	$1,750 +
Middle Right -	$125
Bottom Left -	$150
Bottom Right -	$400 +

Page 185
Top Left -	$100
Top Right -	$100
Middle Left -	$75
Middle Right -	$75
Bottom Left -	$200
Bottom Right -	$75

Page 186
Top Left -	$175
Top Right -	$150
Middle Left -	$1,000 +
Middle Right -	$75
Bottom Left -	$75
Bottom Right -	$300

Page 187
Top Left -	$110
Top Right -	$225
Middle Left -	$180
Middle Right -	$50
Bottom Left -	$50
Bottom Right -	$125

Page 188
Top Left -	$60
Top Right -	$100
Middle Left -	$100
Middle Right -	$250
Bottom Left -	$50
Bottom Right -	$75

Page 189
Top Left -	$100
Top Right -	$225
Middle Left -	$150
Middle Right -	$125
Bottom Left -	$100
Bottom Right -	$100

Page 190
Top Left -	$125
Top Right -	$150
Middle Left -	$125
Middle Right -	$150
Bottom Left -	$50
Bottom Right -	$50

Page 191
Top Left -	$125
Top Right -	$100
Middle Left -	$150
Middle Right -	$125
Bottom Left -	$175
Bottom Right -	$100

Page 192
Top Left -	Rare
Top Right -	$250
Middle Left -	$35
Middle Right -	$60
Bottom Left -	$50
Bottom Right -	$100

Page 193
Top Left -	$100 ea.
Top Right -	$50
Middle Left -	$100
Middle Right -	$300 +
Bottom Left -	$75
Bottom Right -	$150

Page 194
Top Left -	$100
Top Right -	$200
Middle Left -	$300
Middle Right -	$150
Bottom Left -	$150
Bottom Right -	$55

Page 195
Top Left -	$80
Top Right -	$85
Middle Left -	$400
Middle Right -	$70
Bottom Left -	$225
Bottom Right -	$300

Page 196
Top Left -	$150
Top Right -	$125
Middle Left -	$130
Middle Right -	$225
Bottom Left -	$175
Bottom Right -	$225

Page 197
Top Left -	$200
Top Right -	$200
Middle Left -	$175
Middle Right -	$180
Bottom Left -	$250
Bottom Right -	$125

Page 198
Top Left -	$200
Top Right -	$200
Middle Left -	$225
Middle Right -	$100
Bottom Left -	$65
Bottom Right -	$70

PRICE GUIDE

Page 199
Top Left -	$50
Top Right -	$250
Middle Left -	$200
Middle Right -	$225
Bottom Left -	$250
Bottom Right -	$225

Page 200
Top Left -	$600
Top Right -	$250
Middle Left -	$275
Middle Right -	$350
Bottom Left -	$225
Bottom Right -	$200

Page 201
Top Left -	$300
Top Right -	$200
Middle Left -	$200
Middle Right -	$300
Bottom Left -	$120
Bottom Right -	$110

Page 202
Top Left -	$350
Top Right -	$85
Middle Left -	$60
Middle Right -	$100
Bottom Left -	$80
Bottom Right -	$150

Page 203
Top Left -	$200
Top Right -	$80
Middle Left -	$200
Middle Right -	$100

Page 204
$225

Page 205
Top Left -	$750 +
Top Right -	$100
Middle Left -	$200
Middle Right -	$75
Bottom Left -	$150
Bottom Right -	$150

Page 206
Top Left -	$175
Top Right -	$200
Bottom -	Rare

Page 207
Top -	$20
Bottom Left -	$12
Bottom Right -	$20

Page 208
Top -	$12
Bottom Left -	$12
Bottom Right -	$10

Page 209
Top -	$30
Bottom Left -	$35
Bottom Right -	$35